THE GULF MONARCHIES
AND CLIMATE CHANGE

D1740964

MARI LUOMI

The Gulf Monarchies and Climate Change

Abu Dhabi and Qatar in an Era of Natural Unsustainability

CIRS
CENTER FOR
INTERNATIONAL
AND REGIONAL
STUDIES

GEORGETOWN UNIVERSITY
SCHOOL OF FOREIGN SERVICE IN QATAR

Published in Collaboration with
Georgetown University's
Center for International and Regional Studies,
School of Foreign Service in Qatar

HURST & COMPANY, LONDON

First published in the United Kingdom in 2012 by
C. Hurst & Co. (Publishers) Ltd.,
41 Great Russell Street, London, WC1B 3PL
© Mari Luomi
Printed in India

The right of Mari Luomi to be identified as the author of this
publication is asserted by her in accordance with the Copyright,
Designs and Patents Act, 1988.

A Cataloguing-in-Publication data record for this book
is available from the British Library.

ISBN: 978-1849042673

www.hurstblog.co.uk
www.hurstpublishers.com

CONTENTS

CONTENTS

CONTENTS

vii

CONTENTS

LIST OF TABLES

ACKNOWLEDGEMENTS

This book carries two stories; the central one unfolds on the following pages while the other is that of a personal journey that began nearly five years ago in freezing Finland. This journey has been guided and powered by a combination of passion, persistence, curiosity, and coincidence. A number of people have figured importantly in this story.

For the passion that I feel towards the environment and the persistence with which I pursue set goals, I want to thank my wonderful parents. For igniting the spark of curiosity towards the topic of this book, the difficult relationship of the Gulf oil exporters with the environment, I want to thank Dr Tapani Vaahtoranta, my encouraging superior at the Finnish Institute of International Affairs for several years. Finally, it was thanks to my inspirational and supportive doctoral thesis supervisor, Dr Christopher Davidson, at the Institute for Middle Eastern and Islamic Studies of Durham University, that I chose to focus my research on the two dynamic and enigmatic monarchies that form the case studies of this book, Abu Dhabi (and more broadly the United Arab Emirates) and Qatar.

In the past five years, as often happens to people who take on the 'doctoral challenge', this book and my personal life have in many ways become inseparable. There are innumerable people who contributed to this book, either as supervisors, superiors, colleagues, or interviewees, whom I wish to thank. Along the way, many of them have become friends. While I cannot personally acknowledge everyone here, I want to assure that I am equally grateful for your invaluable contributions, which have come in many forms. In particular, for this book, I wish to thank (in alphabetical order): Zahra Babar, Dr Kristian Coates-Ulrichsen, Dr John Crist, Dr Christopher Davidson, Professor Anoush Ehteshami, Laura El-Katiri,

ACKNOWLEDGEMENTS

Wael Hmaidan, Dr Heidi Huuhtanen, Professor Mehran Kamrava, Dr Anna Korppoo, Jim Krane, Ambassador Matti Lassila and Satu Mäki-Lassila, Dr Suzi Mirgani, Dr Lotta Numminen, Dr Mohamed Raouf, David and Liz Roberts, Nadia Talpur, Dr Tapani Vaahtoranta, Dr Antto Vihma, Flora Whitney, and Professor Rodney Wilson.

Also to acknowledge are the several generous institutions that have made this project possible, either through scholarships or by providing a space and a place, with great colleagues, for conducting research. The most important ones are (in chronological order): the Finnish Institute of International Affairs, the School of Government and International Affairs of Durham University, the Foundation of the Finnish Institute in the Middle East, and the Finnish Cultural Foundation. Finally, I wish to offer my most sincere thanks to the Center for International and Regional Studies of the Georgetown University School of Foreign Service in Qatar whose generous Post-Doctoral Fellowship in 2011–12 made possible the transformation of my doctoral thesis into this book.

This journey has brought me to the hot and arid Gulf, far from my origins, via multiple stops, and with many new destinations to come. One person has accompanied and supported me through all these years. It is to Antonio, along with my parents Tarja and Seppo, that I dedicate this book.

ABBREVIATIONS

ADCO	Abu Dhabi Company for Onshore Oil Operations
ADNOC	Abu Dhabi National Oil Company
ADWEA	Abu Dhabi Water and Electricity Authority
ADWEC	Abu Dhabi Water and Electricity Company
CCS	carbon capture and storage
CDM	Clean Development Mechanism (Kyoto Protocol)
DECC	Directorate of Energy and Climate Change (UAE)
EAD	Environment Agency—Abu Dhabi
EIA	environmental impact assessment
ENEC	Emirates Nuclear Energy Corporation
ERWDA	Environmental Research and Wildlife Development Agency
ENGO	environmental non-governmental organization
EWS/WWF	Emirates Wildlife Society in association with WWF
FANR	Federal Authority for Nuclear Regulation (UAE)
FEA	Federal Environment Agency (UAE)
FEWA	Federal Electricity and Water Authority (UAE)
GCC	Gulf Cooperation Council
GSDP	General Secretariat for Development Planning (Qatar)
GTL	gas-to-liquids
HSE	health, safety and environment
IAEA	International Atomic Energy Agency
IEA	International Energy Agency
IISD	International Institute for Sustainable Development
IPCC	Intergovernmental Panel on Climate Change
IRENA	International Renewable Energy Agency

ABBREVIATIONS

LEED	Leadership in Energy and Environmental Design (certification system)
LNG	liquefied natural gas
MENA	Middle East and North Africa
OAPEC	Organization of Arab Petroleum Exporting Countries
OPEC	Organization of the Petroleum Exporting Countries
QF	Qatar Foundation (for Education, Science and Community Development)
QIA	Qatar Investment Authority
QNFSP	Qatar National Food Security Programme
QSAS	Qatar Sustainability Assessment System
QSTP	Qatar Science and Technology Park
SCENR	Supreme Council for the Environment and Natural Reserves (Qatar)
SPC	Supreme Petroleum Council (Abu Dhabi)
UAE	United Arab Emirates
UNFCCC	United Nations Framework Convention on Climate Change

INTRODUCTION

The Gulf monarchies have reached their limits of 'natural sustainability'. The negative consequences of the past decade's fast growth on resource security and environmental sustainability have now become too evident to ignore. Ironically, several of the monarchies have in recent years become dependent on natural gas imports. The scarce water reserves are dwindling, and food import dependence is high. In the past decade, Qatar produced around 50 tonnes of carbon dioxide per inhabitant each year—a dozen times more than the global average. If all people in the world had the lifestyle of an average United Arab Emirates resident, it would require five planet Earths' resources to sustain them.[1] Not only is the Gulf Cooperation Council (GCC) states' environment under stress, but so is the continuation of their status quo of power and wealth. How have the region's governments responded to these challenges? And are the Gulf monarchies' political economies and political systems even compatible with achieving a balance between natural resource use and the environment?

With the economic and population boom in the Gulf monarchies in the 2000s, electricity and water demand skyrocketed, producing unforeseen pressures on domestic energy supply. At the same time, the consolidation of climate change on the international energy agenda created a new element of uncertainty for the region's rulers, whose permanence in power remains largely dependent on fossil fuel revenues, and hence on global demand for oil and natural gas. Simultaneously, the prevailing domestic natural resource consumption patterns have caused unprecedented stress on the region's fragile desert environment. And with the projected impacts of climate change, these existing pressures are likely to intensify. Natural resources, environmental unsustainability, and cli-

1

mate change are now coming together to form some of the GCC states' toughest survival challenges yet.

Fortunately perhaps, the governments of the GCC states are beginning to realize these negative impacts of present development and consumption patterns. The government and institutional responses have—despite many structural similarities among the monarchies of the GCC—come in differing intensities and divergent methods. Many grand promises have been made since the late 2000s, but at the time of writing this book, in 2011–12, only two monarchies had proceeded from mere plans to tangible implementation. Abu Dhabi, the wealthiest emirate of the United Arab Emirates (UAE) and an increasingly self-assertive player, has decided to brand itself as the region's green leader amidst looming domestic energy challenges. Among other things, it has established a multifaceted clean energy company, Masdar, and a civilian nuclear energy programme. Partly because of its federal responsibilities, Abu Dhabi is also trying to find ways to tackle resource consumption patterns at home. Qatar, owner of the world's third largest natural gas reserves, lacks Abu Dhabi's urgency about implementing domestic energy security measures and has opted for a more piecemeal and bottom-up approach to low-carbon development. However, Qatar is quickly becoming a serious regional Research & Development hub for many global academic institutions and companies in the areas of energy and environmental research, and is devising ambitious plans for sustainable agriculture.

At the international level, change and divergence in GCC policy responses are beginning to emerge. Since the early years of international climate negotiations, under the United Nations Framework Convention on Climate Change (UNFCCC) established in 1992, the Gulf monarchies have been predominantly identified with the oil exporters' effort to protect future revenues from international climate change mitigation efforts. The GCC's appearance of unity has stemmed largely from the strength of Saudi Arabia's negotiating strategy, and the absence of strong domestic interests in the smaller states that would clash with this strategy.

Abu Dhabi's and the UAE's passive climate policy began shifting towards more active and independent moderation in 2009. This was caused by a chain of events beginning in 2006, with the establishment of the Masdar Initiative, accelerated by the emirate's victorious campaign to host the new International Renewable Energy Agency in 2009, and culminating in the establishment of a new policy unit under the Ministry of Foreign Affairs in 2010. Qatar, a somewhat more active participant in the UNFCCC, retained an impressively static policy throughout the 1990s and 2000s. In addition to supporting key OPEC (Organization of the Petroleum Exporting Countries) demands, Qatar's position focused on seeking a special recognition for its role as a global provider of natural gas, the cleanest of fossil fuels. Until early 2012, there had been no strong domestic force pushing for progressive engagement in this area.

Focus on natural sustainability

Arguably, the Gulf monarchies' dependence on fossil fuels, fossil fuel revenues, and the social contracts based on these revenues produce unsustainability. This unsustainability is explored in this book through the concept of 'natural sustainability', which is defined as the use of natural resources in a way that allows for prosperity for humans and the environment, at present and in the future. (A further elaboration of the concept is provided in chapter 1.) The book argues that natural unsustainability is a built-in feature of the GCC states' contemporary societies. The current crisis was essentially brought about by the past decade's rapid rise in oil prices globally, and demand for natural resources domestically. It is a direct result of the economic and political dependence on external rent, and partially of the high social dependence on a foreign workforce. Fundamentally, however, the Gulf monarchies' natural unsustainability is determined by three features: the strong 'rentier' elements of their political economies; the persisting authoritarianism of their political systems; and the wasteful ruling bargains in which these two come together. The latter of these has helped to create a counterproductive 'illusion of abundance', the symptoms of which include high and wasteful energy and water consumption patterns and disregard for environmental sustainability.

This study rests on the basic assumption that it is most fundamentally regime survival that drives decision-making in the Gulf monarchies. Policy areas that are crucial for the maintenance of the status quo and the stability of the distributive political economy and more widely of the society, are priority areas for the ruling elites. It is recognized that the GCC states are in many ways still developing countries, and economic growth and security will always remain the governments' top priority,[2] particularly with the Arab uprisings that began in 2011. But environmental sustainability, which has traditionally ranked far below economic sustainability and socioeconomic growth on the GCC governments' agendas, has started to receive increasing attention, which makes it an interesting research topic. As the cases of Abu Dhabi and Qatar demonstrate, fossil fuel wealth can and will be used to promote clean energy and environmental sustainability initiatives if there is an economic or political motive, as in any other state.

With this in mind, the underlying questions that run through this book are: how long can business-as-usual-development continue? Is it possible for a fossil fuel dependent, authoritarian monarchy to adapt to the above-described pressures and survive in the coming decades?

Analytical framework

Departing from a regime survival strategy-oriented approach, this book traces and describes the drivers and motives of change and divergence behind the natural resource and climate change-related perceptions, approaches, and policies of these two monarchies' governments at both domestic and foreign policy levels. In parallel, it examines emerging natural resource-related challenges and vulnerabilities. At the domestic level, the book demonstrates how government responses are produced by the interactions of rentier structures, individual patron figures, the interests and perceptions of the decision-making elite, and the power relationships and dynamics of the decision-making system. This all takes place in constant dialogue with a range of external actors, opportunities and pressures, real and perceived.

At the international level, it analyses the six Gulf monarchies' role in the international negotiations and debates on climate change. Case studies include the GCC states' group dynamics and Abu Dhabi's campaign for hosting the headquarters of the International Renewable Energy Agency (IRENA). These are followed by detailed analyses of Abu Dhabi's/the UAE's and Qatar's UNFCCC policies from the mid-1990s to mid-2012. This comparative approach enables an understanding of the multiple complexities of external climate policy formation. While showing the evident influence of regional and international actors, the book highlights the strong influence of the domestic context, and how the interests and perceptions of the decision-making elite, and the power relationships and dynamics of the decision-making system, are an essential determinant of these external responses.

Positioned in the intersection of Middle East area studies and International Relations, the analysis divides into two levels. The framework used at the domestic level converges with the analytical framework of the external level. For understanding domestic policy choices and developments, the study uses the concepts of rentierism and neotraditionalism, which call attention to the structural dynamics of the GCC states' political economies and decision-making structures, on the one hand; and the key role of the agency of individual top elite members, on the other. These concepts are examined in detail in chapter 1. At the foreign policy level, the theoretical underpinnings of the study (although only elaborated on in this section) are strongly influenced by theoretically pluralist, multi-dimensional explanations, as exemplified in the works of Nonneman and Hinnebusch, among others.[3] The analytical model used in this book has been further expanded and extended to the Middle East regional level in the article 'Gulf of Interest' by the author in 2011.[4]

Case selection

Abu Dhabi and Qatar are among the most structurally similar monarchies in the Gulf. The two are prime examples of a 'strong rentier state', as defined in chapter 1, because of their small national and resident populations, their high GDP per capita, and the large share of external rent in their exports and GDP. Largely stemming from this, both ruling elites also enjoy a high level of leadership autonomy. Despite their small size, Abu Dhabi and Qatar are among the world's top fossil fuel states/entities. Both own substantial oil and gas resources and rank among the world's top exporters, and both are members of OPEC (Abu Dhabi having joined the organization before the creation of the UAE, in 1967). In the recent decade, the two governments have expanded national wealth through record fossil fuel revenues and by accumulating substantial amounts of this external rent in sovereign wealth funds as well as domestic development projects. The modernizing leaderships of Abu Dhabi and Qatar also have ambitious visions regarding the future of their respective emirates. Finally, it should be noted that although Abu Dhabi is not a sovereign state, it acts like one with regard to its local affairs and, increasingly, with regard to many areas of federal and foreign policy-making. This will become evident in the case study chapters 4 and 5, which also incorporate examinations of the relevant federal dynamics that affect

Abu Dhabi's economy and policy-making. In this regard, Abu Dhabi, not to mention Qatar, has actively sought to raise its regional standing and international profile in the recent years.

Research questions and structure of the book

The book begins by asking in chapter 1 why Gulf monarchies have become so unsustainable. It tracks the increasing natural unsustainabilities that are being created in the interactions of the GCC states' political economies, political systems and decision-making structures, and establishes the book's analytical context. Chapter 2 provides a comprehensive analysis of the energy dimension of natural unsustainability. It covers the broadening scope of energy security-related challenges of the six GCC states. Chapters 1 and 2 include detailed empirical examples and quantitative data to support the argument. Chapter 3 turns to the issues of climate change and environmental unsustainability. After an introduction to the international politics of climate change, the chapter explains the two dimensions of vulnerability that the Gulf oil exporters perceive in this matter: climate change itself and international efforts to mitigate it. The chapter also observes GCC states' natural unsustainability through two broadly used quantitative indicators of environmental sustainability, and elaborates on the impact of authoritarianism on the broader societal context through issues like public awareness and non-governmental organizations. Together, chapters 1 to 3 serve to provide a general outline of the three principal new energy and environmental challenges of the GCC: rising domestic energy and natural resource demand; the rise of climate change on the international agenda; and the negative environmental consequences of domestic natural resource consumption and, potentially, climate change.

In the second part of the book, chapters 4 to 7 explore how these multiple natural resource-related pressures have affected Abu Dhabi and Qatar, and how the two governments are responding to them. Owing to the inseparable realities of the federal structure and dynamics, the case study of Abu Dhabi is somewhat expanded to include analysis of the broader UAE, where necessary. The chapters also ask what prompted each monarchy to action. In other words, what were the drivers and motives of change and divergence in Abu Dhabi's and Qatar's responses to the challenges of energy insecurity, climate change and environmental unsustainability? Chapters 4 and 6 provide a broad analysis of the so-called 'natural sustainability complex', or each monarchy's general context of political economy of natural resources, top decision-makers, and structures and dynamics of environmental and climate change-related governance. Chapters 5 and 7 examine the monarchies' responses to climate change and environmental sustainability through case studies of three major alternative energy and sustainability initiatives. They also include a detailed examination of climate change-related resource scarcities and vulnerabilities of each monarchy.

In the third part of the book, chapter 8 examines the GCC states in the international climate regime, with a special emphasis on the evolution of Abu Dhabi/the UAE's and Qatar's perceptions, policies and positions. The chapter tracks the states' positions from the mid-1990s to demonstrate how they have changed (in the case of Abu Dhabi since 2009) or remained stable (in the case of Qatar until 2012), and in which ways the shifting domestic agendas and priorities have impacted on these external agendas. The book concludes with a short reflection on the determinants and successes and failures of the government responses to date, and argues that, despite a promising start in some of the Gulf monarchies, in the absence of profound changes in the existing political economies and state-society configurations they stand a meagre chance of survival in the ongoing century of climate change.

* * *

Owing to the topicality of the issue and lack of previous literature, the case studies in particular rely to a large extent on interviews with UAE and Qatar-based decision-makers, stakeholders and experts. Methodological problems in the fieldwork and research leading to this book have included access to information and statistical data issues. In the former case, access to information and views was complicated by the vertical decision-making structures; lack of established channels of communication and of the habit of dialogue with researchers, in many state institutions; and in many cases self-censorship and fear of engaging in open discussions in areas perceived as potentially sensitive, among others. With regard to statistics, it should be noted that the low quality and often complete lack of reliable statistical data is a major problem for any student of the Gulf monarchies. This is particularly the case, however, with environmental data. Energy and demographic statistics are other problematic areas because of their politicized nature. Arising from this, there are a number of methodological choices to make; in the case of conflicting data produced by local agencies, these all are presented without judgements regarding their validity, whereas in the case of conflicting data from international institutions and local agencies, precedence is given to the former because, and always when, this allows for a comparison between states. Finally, space still remains for further exploration in the Arabic language debates on the issues covered in this book, in particular among nationals of the GCC states.

1

THE GULF MONARCHIES
AND NATURAL UNSUSTAINABILITY

The domestic dynamics of unsustainability

The decade of the 2000s was a period of high economic and population growth in the energy-rich Gulf monarchies. It also introduced a number of new natural resource-related challenges. Some of these emerged from within the states; some were transboundary in nature; and others were produced by the international system. The common denominator is that, if not met with consistent action, these pressures and problems are likely to intensify, hindering development efforts during the coming years and decades.

Domestically, as a result of the rapid growth, demand for energy, water and food in the Gulf monarchies skyrocketed. In many of the states, available supply reserves and mechanisms were put to the test, portending an increasingly insecure and unstable future under business-as-usual development trajectories. Simultaneously, some of the environmental harm caused by accelerating industrial development and urbanization began surfacing. Between 2000 and 2007, electricity generation in the six Gulf Cooperation Council states grew by 65–170% and carbon dioxide emissions increased by an estimated 43–123%. The corresponding OECD averages were 11% and 4%.[1] High dependence on desalination, for between 40% and 99% of all drinking water, acted as a major contributor to energy demand growth, alongside economic diversification and the population boom.[2] Food is another critical area: in 2006–07, the

GCC's total agricultural imports were estimated at US$20 billion. Dependence on the import of basic staples like wheat, rice and sugar in all GCC states except Saudi Arabia is nearly 100%.[3] Pressured between scarce and depleting groundwater resources and rising international food prices, some of the Gulf monarchies began to look for farmland beyond their borders, others sought bold solutions at home.

The Dubai analogy

Altogether, the 2000s was a decade during which the GCC political economies, originally built around plentiful fossil fuel resources and supporting small populations, began to show signs of cracking. The core of the problem, however, was not only the political, economic and social structures created by oil. In these authoritarian political systems, the people on top, primarily concerned about remaining there, also contributed significantly to the course of events. One of these was Dubai's dynamic autocrat Sheikh Mohammed bin Rashid Al Maktoum. Although it was in Dubai's neighbouring emirate Abu Dhabi that some of the first, ambitious proactive responses to the emerging natural resource challenges would be devised, the emblematic 'Dubai model' of the 2000s[4] provides a perfect example of the problems characteristic of that decade's development in the GCC.

Dubai's artificial islands, the Palm Trilogy (Jumeirah, Jebel Ali and Deira) and the World, were devised and partly built over the 2000s on a seemingly endless stream of money. The plans announced in the 2000s boom years were superlative and almost anything was sold as possible: 'the world's largest man-made island'; 'visible from the Moon'. The ultimate aim of Dubai's real estate and tourism strategy was economic diversification, as it was in the other GCC states that soon followed suit. Decision-making in both Dubai and the company behind the Dubai World island projects was highly vertical and largely opaque; both were controlled by the same 'CEO' with a bold vision but without accountability to a 'board', whether directors or citizens. Planning and implementation of the islands were rushed and resulted in unpleasant surprises: in addition to repeated delays to the Palm Jumeirah project, Nakheel, the developer, significantly increased the originally advertised number of villas, and hence decreased space between them, to raise profitability. Later, owing to bad engineering, reports emerged of smelly algae forming in the stagnant seawater between the palm's 'fronds'.[5]

In Dubai, as in the other GCC monarchies, diversification still stood on a weak basis, often depending on fluctuations of the price of oil and international economy. In 2008, following rapid changes in the economic environment, large parts of Dubai's other island projects, including a new addition immodestly titled the Universe, were delayed indefinitely. The same fate befell many other lavish Gulf megaprojects and innovative areas of economic diversification. Most important, economic interests—as was characteristic for both Dubai's man-made islands and development in the GCC in general—coldly overrode considerations of environmental protection and the sustainability of natural resource use. Today, the second palm island in Jebel Ali, with its 70 km of beaches, lies empty, and earlier promises to have 1.7 million people living on and around the island by 2020 have a certain nostalgic air.[6] At the same time, however, despite the economic hardships, energy demand in Dubai, as elsewhere in the GCC, continued to escalate, rising by 10% in 2009–2010.[7] With Dubai's modest oil reserves already near depletion, the government began contemplating new supply side solutions. Rather counter-intuitively, the emirate announced in 2011 that coal would feature prominently in its mid-term energy mix, in addition to natural gas, nuclear power, and tiny quantities of solar energy.

Undoubtedly, as promised by their constructors, both Dubai's mega-projects, and the 2000s development in the Gulf monarchies more broadly, will make history. As environmental awareness in the region and globally rises, however, they may be remembered more for their profound impact on their natural surroundings than as showpieces of successful man-versus-nature development. The palm island of Jebel Ali, for example, stands on top of a marine sanctuary, once home to fish, coral and sea-grass beds. Dubai's coal plants, if built, may go down in history as symbols of an oil monarchy that lacked the ambition to look beyond the unsustainable development trajectories of the industrialized world. The 2000s' development in the Gulf monarchies will be remembered by future generations. As this chapter suggests, this will not be for the reasons our contemporary monarchs would desire, but rather as an era of badly managed and short-sighted, naturally unsustainable growth.

New transboundary pressures and the rentier legacy

The emirate of Dubai represents one extreme in the GCC states' oil wealth continuum: with its oil production peaking in 1991,[8] Dubai became a 'declining oil rentier state' in which the rulers had only one choice for securing their survival: to successfully diversify the economy away from oil. The least restrictive path for the local elites' autonomy would be diversification into other sources of external rent, which would allow elites to maintain the existing mechanisms of co-option of potential sources of discontent and opposition. Even this would still be conditional on two factors: the citizens' expectations not exceeding the state's capacity to allocate, and the state being able to provide the citizens with meaningful jobs. As another option, the government could still seek to steer diversification into non-rent sources of income. In this scenario, however, control over massive financial resources would shift away from the top, and the rulers would inevitably face a choice between meaningful democratization and survival through coercion. While this was the classical rentier-reform interpretation, the unravelling of the Arab Spring in 2011 demonstrated that, although in wealthier states of the Arab World internal calls for reform seemed less likely, nothing was to be taken for granted any more. Prompted by uprisings in Tunisia, Egypt and elsewhere, the Arab Spring reverberated practically throughout the entire Arabian Peninsula. Signs of political awakening and discontent were most visible in the declining rentier monarchies of Bahrain and Oman, but were also felt in Saudi Arabia, Kuwait and the poorer parts of the United Arab Emirates.

On the surface, only two monarchies remained relatively stable and quiet: Qatar and Abu Dhabi. These two, the latter being part of a seven-emirate (con)federation, represented one extreme of Gulf monarchies' wealth. Measured by dependence on oil and natural gas revenue, per capita wealth, and elite autonomy, Abu Dhabi and Qatar are currently the strongest and richest rentier states not only in the Gulf, but in the entire Middle East and Africa region. In the 2000s, supported by record revenues in the oil and gas sector, small national populations and dynamic leaderships with wide autonomy over financial resources and decision-making, they quickly rose from regional backwaters to dynamic regional actors boldly competing with and challenging their neighbours, from small to large, in areas as diverse as education, overseas investment, finance, real estate, tourism, and defence and foreign policy. In hindsight, it was no major surprise that some of the region's most innovative approaches to alternative energy and environmental sustainability would emerge from these two monarchies. The big question that remained was:

will the existing political systems ever be capable of bringing about a transformation into sustainable, low-carbon knowledge economies?

An even further-reaching transboundary force that threatens the Gulf monarchies' socio-political configurations is climate change. In the late 2000s, as a result of mounting scientific evidence and increasing public awareness, the need to avoid dangerous climate change (a global temperature increase of over 2°C)[9] started to figure as a key priority in the global energy debates. The breakthrough of climate change on the international agenda coincided and converged with the Gulf monarchies' intensifying domestic natural resource-related pressures. The region's people and governments had previously given equally little attention to either. This was for two reasons: the international climate change negotiations traditionally functioned as an arena for the OPEC member states to defend their fossil fuel-revenue dependent economies against the impacts of international mitigation efforts; and local resource consumption patterns and progressing environmental degradation in the Gulf received attention only from a few, largely Western-educated professionals, often working in the region temporarily, and disconnected from each other. The dominant concern of most nationals, expatriates and governments was sustaining economic growth. Sustaining the environment was no-one's business.

The challenges of climate change are not unique to the Gulf: rising temperatures and sea-levels threaten countries world-wide, but the hot and water-scarce regions of the Middle East and Africa in particular are expected to suffer. For the moment, the Gulf monarchies, which are on average much wealthier than their neighbours, are better placed to adapt to the physical challenges of climate change, for example by elevating flood barriers and paying for an increased need for air conditioning and desalinated water. However, the high economic dependence on fossil fuel exports of the Gulf monarchies means that these states need to pay special attention to the international politics of climate change.

In addition to the potential negative consequences of climate change, those Gulf monarchies that still own large reserves of oil and natural gas—Saudi Arabia, Kuwait, Abu Dhabi and Qatar—also face potential negative consequences from the nascent global shift to low-carbon economy. The aim of international climate change mitigation, to drastically cut global greenhouse gas emissions (without viable decarbonizing technologies), implies a move away from the oil and natural gas that simultaneously function as the Gulf monarchies' main exploitable natural resources, principal exports, and key internal stability resources for the local ruling elites. The intra-GCC differences in oil and natural gas wealth laid out above present a dualistic picture here too: while the declining rentiers of the Arabian Peninsula, principally Bahrain and Oman, are struggling with their domestic natural resource pressures and the need to diversify rapidly away from oil rent dependence, the economic stability of the more robust rentier states in turn is guaranteed for the time being, but only as long as global demand for fossil fuels is sustained. Evidence of the diversity among the Gulf monarchies, usually portrayed as a monolithic group, is the fact that in material terms these now stand on different sides of the 'peak oil' concept: some struggling with a domestic supply peak, others fearing a global demand peak.

Together, the rising internal, transboundary and external pressures and challenges discussed above threaten the long-term prosperity of the Gulf monarchies' political economies. They also call into question the future of the unwritten social contracts, or 'ruling bargains', between the local rulers and Gulf nationals. This and the following chapters will demonstrate how the period starting from the early 2000s became one of economic and environmental pressures and unsustainability and brought the GCC monarchies to a critical point, at which a profound rethink of the existing economic-political models has arguably become a necessity. Nevertheless, there is still great variation in perceptions among Gulf monarchs and populations regarding the urgency to act.

Since the twelve unique monarchies of the Gulf (including the seven emirates of the UAE) have been faced with different mixes of these same pressures, they have also reacted and responded in somewhat diverging ways, according to their respective economic capabilities and varying domestic preferences. As will be demonstrated, the GCC governments' early responses to the three interlinked challenges of energy security, climate change and environmental sustainability have been produced by complex interactions between the structures of political economy and interpretations of national and personal interest by top elite members, and shaped by the dynamics of local decision-making systems. This chapter specifically concentrates on the first aspect of this triad, the Gulf monarchies' political economy, and its reproduction of what is termed in this book as 'natural unsustainability'. Chapters 2 and 3 examine how this type of political economy has interacted with the issue of domestic natural resource consumption and how its basis is being challenged from the outside, as a result of major shifts on the international energy agenda. Chapter 3 also surveys the environmental unsustainabilities produced by the 'naturally unsustainable development' of the 2000s, and shows how climate change is expected to act as a challenge intensifier, both increasing and complicating existing challenges in this area.

The political economy of natural unsustainability

The multiple meanings and political exploitation of the terms 'sustainable development' and 'security' have arguably diluted their usefulness as analytical concepts. Similarly, the word 'sustainability' can have almost opposite meanings when attached to notions like economic growth or the Earth's carrying capacity. While concepts like 'environmental sustainability', 'environmental security', and 'natural security', the last coined by Burke,[10] are useful for emphasizing certain priorities and areas of the *problématique*, the authoritarian Gulf monarchies of the 2000s and 2010s are arguably not in need of more securitization. Nor is the extended concept of security helpful for describing the broader picture painted here, which is that of limited natural resources and balancing between their internal and external consumption. The Gulf monarchies' environmental sustainability is eroding fast, but this is the symptom, not the source of the problem. Hence the need for a new concept: 'natural sustainability'.

A naturally sustainable state, as argued in this book, is one in which the valuable natural assets of a state, both renewable and non-renewable, are consumed in a way that allows for prosperity for both present and future generations while achieving a balanced relationship with the surrounding environment (that is, environmental sustainability). It is therefore in tune with the most common definition of sustainable development by the Brundtland Commission in 1987: 'meeting the needs of the present without compromising the ability of future generations to meet their own needs'.[11] Since human societies keep evolving (and should do so), natural sustainability refers simultaneously to a process and a state of affairs. While natural sustainability does not necessarily require economic growth, economic and natural sustainability go hand in hand. (It should be stressed that economic growth and economic sustainability are not synonymous.) Expanded notions of security, such as energy, food, water and environmental security, are closely related, as a naturally sustainable political economy is also more likely to achieve security in these key areas. Because of its focus on natural resources and their consumption patterns, the concept of natural sustainability is also particularly well suited for the incorporation of the study of the politics and policies relating to climate change and the evolving science around it. Climate change is an environmental problem, caused by unsustainable natural resource consumption patterns. Technologies for its mitigation are already available, which makes its solution principally an economic and political one. Hence the need to examine both the political economy and decision-making process and outcomes in a state or political entity, so as to understand what drives the sustainabilities and unsustainabilities of respective development pathways.

Finally, to clarify the ontological foundations of this book: its argument is based on a relatively ecocentric understanding of humanity as a part of the Earth's ecosystem. According to this view, human development should be constantly nurtured and improved, but this should not be done at the expense of the environment. Environmental sustainability, it is believed, has an intrinsic value, and is not only valuable for what it can contribute to human prosperity. While this book analyses the ways in which natural unsustainability can make entire economies, polities and societies in the Gulf insecure and unstable, and examines some of the recent attempts to fix the problem, the author believes that the ultimate value of achieving natural sustainability goes even beyond these single-generational issues.

The building blocks of GCC political economies

The 21st-century Gulf monarchies are still essentially 'rentier states' underpinned by an intricate, unwritten social contract between the rulers and the citizens, under which the government allocates a large share of the external rent to its citizens who in exchange surrender a large share of their demands for political rights. Of course the level of rentier wealth and leadership survival strategies varies from country to country. The idea of rentierism and its origin are familiar to students of the region but, as a short introduction, it should be noted that it has been commonly used as an analytical tool for explaining the nature and dynamics of Gulf monarchies: their emergence, the character of their political systems, the persistence of authoritarianism despite modernization, and their policies and policy-making processes. Academic debates around rentierism have considerably evolved, with the conceptual idea gaining new dimensions, conditionalities and depth through the works of Herb, Chaudry, Davidson and others, undergoing what Gray defines as three phases (classical; specialized and conditional; and late rentierism).[12] It has been increasingly challenged by authors based in or coming from the region, like Hertog and Abdullah,[13] who argue that the surrounding debate lacks empirical analysis, enforces a notion of the region's exceptionalism, and paints a static view that oversimplifies the dynamics and potential for change of the GCC monarchies and their state-citizen-relationships. Nevertheless, arguably the concept still holds undeniable, almost paradigmatic explanatory power for structural analyses of the surprisingly resistant authoritarian regimes of the Gulf. Rentierism in a simplified form is a macro-level structural concept in that it contributes to understanding of the implications of oil wealth for the distribution of power among rulers and citizens. It therefore manages to tell little of the influence of personal preferences, interests and choices of individual leaders, and not much of the characteristics and dynamics of state institutions. Despite this, as will be shown in this book, the era of the rentier state is far from over and the concept still remains superior in explaining key aspects of the Gulf monarchies' contemporary natural resource problems.

The classical definition of a rentier state was put forward by Mahdavy in 1970: it is a country that receives 'on a regular basis substantial [a] mounts of external rent' from external actors,[14] while Beblawi's widely used definition of rent is 'reward for ownership of all natural resources'.[15]

Writing in 1970 on Iran's economic development in the 1950s, Mahdavy observed that the ability of the rentier governments to avoid taxation translates into significant independence of their citizens, but that despite the associated capabilities to both bribe and coerce individuals and groups, the rulers' power remains always highly vulnerable owing to its dependence on external rent.[16] This still remains the essence of the concept of rentierism. Beblawi's contribution to the rentier state concept included three conditions: the predominance of external rent in the income of the economy, often paired with a weak productive domestic sector; a situation in which only a minority of the population is engaged in generating this rent while the majority is involved exclusively in either distributing or using it; and the role of the government as the principal recipient and distributor of this rent.[17] Hence Luciani's parallel term 'allocation state', which he defined as a state that derives over 40% of its revenue from oil or other foreign sources and has expenditure that corresponds to a substantial share of GDP. An allocation state's main function is the distribution of rent, while in 'production states' the state engages in both production and reallocation.[18]

In the late 2000s, both post and pre-crisis, all the GCC states fitted both Beblawi's and Luciani's definitions of rentier/allocation state. The most important structural variation among the GCC states is found in the size of their economies, as table 1 shows. Owing to the UAE's loose federal structure, which grants its wealthiest members pronounced autonomy in local decision-making, and the diverging economic bases of its two largest emirates, Abu Dhabi and Dubai are included in a number of tables in the chapter.

Two key factors have defined the development and survival of the rentier monarchies in the Gulf: in a more historical perspective and in the case of the smaller monarchies, the role of Britain as a protector of the

Table 1: Indicators of rentierism in the Gulf monarchies.[19]

	Share of fuel exports of merchandise exports (2008–09)	Share of oil rents of GDP (2008–09)	Share of fossil fuel revenue of government revenue (2008)	GDP (US$m, 2009 [global ranking])
Bahrain	69%	22%	85%	20,595 [97]
Kuwait	95%	60%	77%	109,463 [55]
Oman	83%	37%	87%	46,866 [72]
Qatar	83%	19%	57%	98,313 [56]
Saudi Arabia	89%	54%	89%	(2010) 434,666 [22]
United Arab Emirates	65%	24%	80%	(2010) 297,648 [31]
– Abu Dhabi	(2010) 72%	*(2010) 50%	*(2010) 83%	(2010) 168,890 [n/a]
– Dubai	–	(2007) 5%	–	(2010) 110,090 [n/a]

* Only oil revenues.

local rulers in the pre-oil 20th century,[20] and more recently, the magnitude of oil wealth in relation to the size of the citizen populations (see table 2). Despite high national population growth in the past decades, relatively high external rent has enabled the rulers of the wealthiest monarchies to assert their power through the inclusion of most nationals in small rentier 'elites', with special economic and social privileges, while expatriates are excluded from most of the rights and benefits enjoyed by the nationals, including citizenship. Important differences, however, exist in the level of benefits enjoyed by the citizens in the different monarchies.

One way to demonstrate the levels of traditional rentier wealth and the strength and durability that the rentier state still has (provided that international demand for oil and gas is sustained, and not taking into account domestic consumption) is to calculate the number of barrels of oil presumed to be left in the ground per citizen. As shown in table 2, there is vast variation: from 230 barrels in Bahrain to over 100,000 barrels in Qatar and the UAE. In Qatar, on top of this come the country's massive natural gas reserves, which—given its small national population—translate to roughly 115 million cubic metres of natural gas (over 700,000 barrels of oil equivalent) per citizen in 2010, making Qataris by this measure the richest people in the world. Differentiation can also be observed through rentier wealth per capita: according to the Qatar National Bank, in 2009, the country's oil and gas revenue per national was US$79,000; significantly higher than in any other GCC state. The closest runners up were the UAE (US$46,000) and Kuwait (US$41,000). In the three remaining states, fossil fuel rent per national was only US$6,000–7,000 that year.[21]

The GCC monarchies currently display rather diverging demographic makeups, with the estimated share of nationals ranging between 5% in Dubai and around 70% in Oman and Saudi Arabia, as shown in table 2. Lower GDP per capita roughly correlates with lower levels of foreign workforce, as in the less wealthy states like Saudi Arabia, Bahrain and Oman a part of the citizen population is compelled to accept lower-income jobs and work in the private sector. Owing to the exclusive rent allocation structures and the wide income disparities, particularly in the wealthiest monarchies of Abu Dhabi and Qatar, economy-wide indicators such as per capita GDP are not very useful; a national's average income and that of an low-skilled Asian worker in these societies are two worlds apart (and further analysis is complicated by the lack of reliable Gini coef-

Table 2: Levels of rentier population and wealth in the Gulf monarchies.[22]

	Total population (millions, est.)	National population (est. share of total)	National workforce of total workforce (est.)	GDP per capita, PPP (current intl. $)**	Remaining proven oil reserves per national (barrels)
Bahrain	(2008) 1.107	(2008) 49%	(2008) 23%	(2008) 25,799	(2008/11) 230
Kuwait	(2009) 3.485	(2009) 32%	(2009) 17%	(2007) 52,657	(2009) 91,015
Oman	(2008) 2.867	(2008) 69%	(2008) 25%	(2009) 26,791	(2008) 2,831
Qatar	(2010) 1.697	(2010) 13%	(2010) 6%	(2009) 80,944	(2010) 117,401
Saudi Arabia	(2008) 24.807	(2008) 73%	(2008) 49%	(2010) 22,713	(2008) 14,584
UAE	(2008) 4.765	(2008) 19%	(2010) 15%	(2010) 47,215	(2008) 108,025
– Abu Dhabi	(2010) 1.968	(2010) 22%	(2008) 11%	*(2010) 85,845	(2010) 212,546
– Dubai	(2010) 1.905	(2007) 5%	–	*(2007) 41,652	–

* Not PPP-adjusted.
** Purchasing power parity.

ficient data). Similarly, the wealth disparities and the massive presence of energy and heavy industries render the notion of 'a' per capita CO_2 emission problematic and highly politicized, as will be explained later.

The socio-economic consequences of the above-described structural realities are numerous. The absence of a need to tax citizens leads to a lack of ruler accountability and hence reinforces the political system's authoritarianism. As indicated by international comparisons, despite the past two decades of 'political decompression'[23] through quasi-reforms and limited openings the GCC states continue to remain far from democratic (see table 3). The atomization of society along ethnic and income-based lines and the quasi-absence of autonomous civil society institutions have been further consequences.[24] The latter is visible in the small and toothless environmental civil societies of the GCC states. Furthermore, the line between private (in terms of wealth or interest, for example) and public (in terms of funds or service) is often blurred as members of the political elite engage in private sector-like business activities, such as the majority stake of the emirate's ruler in Dubai World, and civil society-type organizations, like the Qatar Foundation, led by the Emir's wife. Another feature is the government's function as the major employer for the citizens: from 34% in Bahrain to 87% in Qatar (see table 3). This practice has led to high expectations regarding employment, salary and career advancement, and also to decreased efficiency in the public sector, as employment is often seen as a granted benefit rather than something achieved.

Furthermore, at the level of individuals, the concept of 'rentier mentality' offers essential observations of collective behaviour patterns in rentier states. The implications of the 'break in the work-reward causation'[25] for work ethics have become a key characteristic of the darker side of the Gulf citizen stereotype, but rentier mentality also affects overall populations' natural resource consumption patterns through the 'break in the cost-consumption causation', as will be explained later.

Typically, most oil-exporting states, despite varying characteristics, have suffered from political and economic crises and seem to be locked in Karl's paradox of plenty. Oil rents create unstable political economies and produce a barrier to change because the required political reforms are not in the interests of the political leaders. Although the small Gulf monarchies have mostly managed this inherent instability (either through exclusive rentier bargains or with the support of other rentier regimes,

Table 3: Political and labour market implications of rentierism in the Gulf monarchies.[26]

	Level of authoritarianism		Share of public sector employment of total employment of nationals
	*Economist Intelligence Unit**	*Freedom House***	
Bahrain	3.49 (122/167)—authoritarian	5.5—not free	(2010) 34%
Kuwait	3.88 (114/167)—authoritarian	4.0—partly free	(2008) 86%
Oman	2.86 (143/167)—authoritarian	5.5—not free	(2008) 47%
Qatar	3.09 (137/167)—authoritarian	5.5—not free	(2009) 87%
Saudi Arabia	1.84 (160/167)—authoritarian	6.5—not free	(2008) 72%
United Arab Emirates	2.52 (148/167)—authoritarian	5.5—not free	Abu Dhabi: ***(2008) 63%

* Scores range from 10 to 0, with higher scores indicating higher levels of democracy.

** Scores range from 1 to 7, with scores over 5.5 denoting 'not free'.

*** Public administration and defence only.

as in the case of Bahrain in March 2011), they share a crucial characteristic with other 'petro-states' in that their 'framework of decision-making [is] both constructed and subsequently based upon highly politicised
allocation of rents'. This creates a perverse incentive structure, characterized by the postponement of needed structural changes, institutional
rigidity, lack of policy innovation, and 'a policy style marked by an exaggerated tendency to throw money at problems'.[27] Despite massive efforts
and improvement, the GCC states have at least partly fallen victim to
the resource curse: economic diversification in the six states has not radically advanced from the 1970s, and economic growth has remained
dependent on and vulnerable to fluctuations in oil prices. In the five years
before the onset of the economic crisis in 2008, GDP growth in the six
GCC states averaged 8.3% per year.[28] Growth was particularly fast in
gas-rich Qatar: 13.5% per year in 2001–09. After the crash in oil prices,
from US$147 per barrel in July 2008 to US$32 in December that year,
GDP growth in the GCC was reduced dramatically, only showing signs
of recovery from 2010 onwards, with the recovery of oil prices.

The global economic downturn that started in late 2008 was also a stark
reminder of the vulnerability of attempts at diversification into non-oil
sources of rent. The case of the most globalized monarchy, Dubai, is a case
in point. Already since the 1970s, but increasingly since the 2000s oil
price hike, governments of the Gulf monarchies have accumulated rent
in sovereign wealth funds to provide financial buffers over periods of low
or negative growth. By late 2011, they had secured estimated total assets
of close to US$1.6 trillion.[29] The global financial crisis and the simulta-
neous decline in oil prices it caused are estimated to have made a large
dent in both the oil revenue and the sovereign assets of Abu Dhabi, Qatar
and Kuwait (with sovereign wealth losses alone estimated at 12–41% in
2007–08).[30] Nevertheless, it was Dubai, hailed as the most diversified of
the Gulf economies, that got the worst hit due to a collapse in real estate,
tourism, shipping and the global economy in general. Still, Dubai's future
strength may prove to be precisely its low dependence on oil revenues.[31]

Regime legitimacy strategies and the importance of human agency

In the past, tribal relations were a collective force that constituted a major
challenge to the Gulf Arab rulers, with the successful ones being those
that 'tamed' this force.[32] Indeed, while rentier structures influence eco-

nomic, political and social outcomes, the importance of individual rulers and top elite members should not be downplayed either. Since the onset of fossil fuel revenue, the outreach of the GCC monarchs' rule has been vastly extended and they have been able to create allocation states and manipulate societal relations by supporting, co-opting and coercing groups and individuals.

In addition to skilfully employing the available financial resources for survival, Gulf ruling elites have simultaneously employed a range of immaterial strategies aimed at increasing domestic legitimacy, often with reference to historic continuities. As modernization revisionists like Hudson, Nonneman and Davidson have argued, alongside rentier structures the skilful exploitation of traditional resources also plays a role in explaining the longevity of monarchy in the Gulf. Despite their apparent authoritarianism, the monarchies of the Gulf have maintained traditional forms of interaction and channels of input (such as the ruler's *majlis*), a level of social pluralism, and even something akin to a civil society, although all these have mostly been co-opted and their powers strictly limited.[33] Anderson speaks of 'official civic myths', to which Gulf monarchies have devoted significant time and resources since their independence.[34] In addition to traditional authority, Gulf ruling families have also sought charismatic authority. In the case of Abu Dhabi and the UAE, the legend of Sheikh Zayed bin Sultan Al Nahyan as the father of the nation and as a conservationist has arguably been created and recreated throughout the past decades for the purpose of upholding the domestic legitimacy of the patriarch figure and his sons.

Rather than countering modernity, the monarchies in the Arabian Peninsula have increased their legitimacy by materially embracing it. The oil wealth and consequent external technical assistance have also played an important role in building governments' strength and capabilities.[35] Qatar under its current Emir, Sheikh Hamad bin Khalifa Al Thani, is an excellent example of a modernizing monarchy that has cleverly exploited its natural gas resources and the appearance of institutional reform and quasi-non-governmental organizations to strengthen the ruler's power base and consolidate its political legitimacy, as Kamrava has observed.[36]

At the same time, Gulf monarchs also seek to maintain the existing, strong vertical power relationships. These methods are best described by Weber's term 'neopatrimonialism', which he characterizes as 'new ways of using what only superficially resembles the old patrimonial style and mechanisms'.[37] As explicated by Davidson, instead of resorting to the survival strategies described by Huntington for the 'king's dilemma'— voluntary transformation of the polity, institutionalized coexistence, or resisting reform—the rulers of the UAE, for example, have created a society-wide extension of the ruler's personal network in which all links are tied to the top. In addition to culture, religion, personal resources and charisma, Davidson names other types of resources and strategies of power legitimization, including ideology, local identity, and institutionalization of authority.[38] In the case of Abu Dhabi, and to some extent the UAE, environmentalism has served as a legitimacy resource, as will be demonstrated.[39]

Rentier structures do not inevitably create natural resource and environmental challenges. Neither are government responses to these challenges mechanically caused by the rentier structures. Rather, the human element, or the top decision-makers, can influence trajectories and outcomes by virtue of the financial assets and political autonomy granted not only by external rent but by the role that each decision-maker manages to carve for him- or herself. Similarly to CEOs of large companies, individual Gulf top elite members wield decisive influence in major strategic decisions.

This book argues that the legitimization strategies and structures described above, together with the overarching rentier state structures, form the basis for understanding the small Gulf monarchies' domestic context. It is in this context that foreign policy interests and, consequently, international climate change policies are also largely shaped. Decision-making structures can also influence which issues receive the government's attention and whether plans and policies are followed through. Apart from institutional size and capacity, important factors to take into account include leadership patronage; the level of establishment, scope and impact of activities of an institution; inter-institutional relationships, including turf demarcation; and the influence of other governmental and non-governmental institutions.

The resource triad and influence of the human factor

In a nutshell, the rising unsustainability of the Gulf monarchies is due to a triad of key 'natural' resources: energy, water and human beings. In the case of energy, the problem is that there is too much of it. Fossil fuels constitute central components of both energy security and economic sustainability. Oil and natural gas abundance has become inseparably interconnected with the political legitimization and survival strategies of the Gulf ruling elites. This has formed a key driver of the rampant growth in domestic energy demand. The abundance of domestic energy and the authoritarian welfare states devised around it have created wasteful consumption practices and mentalities, and structural inefficiencies. Energy subsidies, or the practice of selling for less than the market value, have come to form practically insurmountable barriers to energy efficiency, to using cleaner energy sources, or to incentives for energy and water conservation—whether in the residential or industrial sector.

The six GCC states are the epicentre of the world's fossil fuel supply: together they own 36% and 22% of proven global oil and natural gas reserves and account for 21% and 9% of total global production, respectively (see table 4). In past decades, smaller populations and lower levels of socioeconomic development translated into lower domestic energy demand and enabled the establishment of natural resource subsidy regimes. These subsidy regimes still exist today in the form of cheap petrol and diesel and cheap or free electricity and water. In 2010, the price at the pump in the six states was only US$0.15–0.44 per litre of petrol, compared for example with US$1.80 in the UK and US$0.70 in the US.[40] Notably, the large variation in proven remaining oil reserve life, ranging from roughly a decade in Bahrain to over 100 years in Saudi Arabia, is not reflected in petrol prices, as is shown in table 4. The fact that prices in 2010 were the highest in the UAE, with a long reserve life, is explained by the lack of oil in all other emirates except Abu Dhabi, and a federally set cap on prices.

As for water, the problem is that there is too little of it. All six GCC states suffer from absolute water scarcity, defined as less than 500 m^3 of available renewable water resources per capita per year. The situation is particularly acute in Kuwait, with the country's annual renewable water resources at 20 million m^3, equal to a mere 5.7 m^3 per resident per year for a population of roughly 3.5 million in 2009.[41] However, the abundance of cheap energy and the availability of desalination technologies distort the way in which water availability is perceived. Desalination creates the illusion of water being as abundant as oil or natural gas. This might be the case from a technical perspective, but this artificial abundance is achieved at high cost to the environment, through greenhouse gas emissions and highly saline, chemical-rich discharges. Water is also produced at a high opportunity cost, through exportable energy resources consumed domestically. Furthermore, there are important linkages between food security and water scarcity, as 'virtual water' acquired through food imports, alongside desalination of drinking water, is the paramount enabler of life in the region.

Table 4: Indicators and symptoms of energy abundance in the GCC.[42]

	Proven oil/gas reserves (of global total, end 2010)	Oil/gas production (of global total, 2010)	Oil reserve/ production ratio (years, 2010)	Petrol prices (US$/litre, 2010)
Bahrain*	Negligible	0.2/0.4%	**7	0.27
Kuwait	7.3/1.0%	3.1/0.4%	100+	0.26
Oman	0.5/0.4%	1.0/0.8%	17	0.41
Qatar	1.9/13.5%	1.7/3.6%	45	0.19
Saudi Arabia	19.1/4.3%	12.0/2.6%	72	0.15
United Arab Emirates	7.1/3.2%	3.3/1.6%	94	0.44
– Abu Dhabi	6.7/3.0%	–	–	–

* In addition to territorial reserves, Bahrain shares the output of an oilfield with Saudi Arabia.
** Data for 2009.

The third critical resource is human beings. Humans are not generally regarded as a natural resource, but rather as the users of natural resources. Nevertheless, from the perspective of a government, humans are a resource that needs to be managed, although under different rules and norms. Although this is a contentious topic and categorization, there is arguably too much *and* too little of this resource, from a sustainability perspective. The last decade's economic growth and diversification drive increased the general need for labour. The persisting problem of a mismatch between skills, expectations and morale among national workforces in many of the monarchies fed into the need to import this labour. The major problem, from this study's point of view, is that GCC nationals and expatriates have a collective mentality that produces unsustainability. This dimension of the human resource element is less tangible, but is at the heart of the region's sustained natural unsustainability: most of the Gulf monarchies' population simply does not think about the environmental consequences of their lifestyles and everyday choices. Reasons for this range from poverty to lack of education and different cultural backgrounds; lack of financial, and infrastructural and legal incentives to curb overconsumption, waste and pollution; and *nouveau riche*-type lifestyles that emphasize Western consumerism and materialism. In this setting, leading change at the individual level becomes complicated by overwhelming everyday obstacles, including weak public transport, weak recycling infrastructure and fossil fuel-dominated electricity markets. Small personal actions seem futile, which may lead even the most eco-conscious individuals to feel indifferent. Human resource-related unsustainability in the GCC states is therefore both quantitative and qualitative.

Beginning with the quantitative aspect, from a system stability perspective, population growth, as it interacts with the local rentier structures and social contracts, has become a key source (alongside economic growth) of natural unsustainability in the Gulf monarchies. It impacts negatively on the per capita availability of both energy and water resources and allocable wealth. In a tragically Malthusian way, population growth also bears important consequences for food security, particularly in countries highly dependent on imports. Population growth, and in particular that of the national population, also threatens the continuation of the allocative state system and the power of the ruling elite, given GCC citizens' expectations of sustained or rising welfare benefits.

As an indirect result of the boom in oil prices since 2002, the Gulf monarchies attracted increasing numbers of immigrants. Economic and population growth largely correlated throughout the decade: Qatar and Abu Dhabi, with double-digit GDP growth, also attracted on a relative basis the most immigrants. In this period, the two emirates' population grew by an average of 14% and 10% per year, respectively, according to official statistics (see also table 5).[43] Partly overwhelmed by the large influx of immigrants, partly as a result of failing labour force nationalization and job creation policies, the smaller Gulf monarchies began struggling with demographic imbalances, which became magnified and hindered economic and social development efforts. The imbalances include dependence on a highly mobile foreign workforce, particularly in the private sector; an unproductive and expensive public sector employing nationals with high salaries (while simultaneously relying on foreign professionals and consultants); high unemployment among nationals, particularly women; and high expectations regarding employment and career advancement. Perhaps most worrisome is the GCC's low-performing education sector. Worsening future prospects is the fact that, since most of these problems are attributable to the GCC allocation state and the excessive reliance on external rent for economic development,[44] they are therefore likely to persist as long as the existing ruling bargains are maintained.

Table 5: Population and economic growth in the Gulf monarchies.[45]

	Total population in 2000 (World Bank est., millions)	Total population in 2010 (World Bank est., millions)	Population growth 2000–10 (average/a, est., 2000–10)	Economic growth 2000–10* (average/a, 2000–10)
Bahrain	0.638	1.262	7.1%	6.3%
Kuwait	1.941	2.737	3.5%	7.0%
Oman	2.264	2.782	2.1%	4.9%
Qatar	0.591	1.759	11.5%	13.5%
Saudi Arabia	20.045	27.448	3.2%	3.4%
United Arab Emirates	3.033	7.512	9.5%	5.8%

* Data not available for all years for all monarchies.

For several decades, the GCC governments have employed strategies to safeguard the allocative system and the state-citizen rentier bargain. These include limited citizenship pathways; encouraging larger family size among nationals through welfare benefits and other subsidies; and retaining the ability to manipulate the share of non-nationals through the *kafala* or sponsorship system. However, the segmentation of the labour market and the infrastructure, and real estate-heavy routes to economic diversification, effectively nullify the states' ability to curb the demographic imbalance, which is led by the influx of construction and service sector employees from Asia and the Middle East.

The labour nationalization policies that started in the 1980s,[46] aimed at increasing the share of nationals in the private sector, have been largely unsuccessful. This is shown in table 2: some 51–94% of the GCC states' labour forces in 2008–10 consisted of non-nationals. Government is still a major employer of nationals: in the mid-2000s, government salaries made up over 10% of the GDP in most GCC economies. Other reasons for failure include the lack of functionality (or producing the right skills and attitudes) of the educational systems, and the high salary expectations and rentier mentality that have eroded the work ethic of nationals.[47] As GCC nationals' population pyramids remain bottom-heavy, the pressure on governments to succeed in large-scale job creation for citizens is growing by the day. As a result, the employment of nationals figures as a core element in the medium and long-term plans of most GCC monarchies, in particular after the onset of the Arab Spring.[48]

GCC populations are still very young. Even with the accelerating population influx consisting predominantly of single adult men, according to World Bank estimates, in 2010 the share of population of ages 0–14 in the total GCC populations was 13–30%. The highest shares of youth (27–30%) were registered in states with the highest shares of nationals in the total population, Oman and Saudi Arabia.[49] The UAE census of 2005 calculated the share of Emiratis under 20 at over 50% of the total national population.[50] As an indicator of how GCC governments are struggling with providing these new generations employment, a study from 2004 estimated youth unemployment as high as 49% in Oman, 26% in Saudi Arabia, 21% in Bahrain and 12% in Qatar, where 50% of young women were unemployed.[51] A continuing failure to create meaningful employment for the growing national population is not only detrimental to the economy but also a concern for the local regimes, as unemployment renders citizens unsatisfied. Here, economic diversification and labour nationalization are linked directly to political stability, particularly in the weaker rentier monarchies where falling per capita rent levels dilute local ruling bargains to a level that no longer serves as a salary substitute.

Furthermore, apart from the subjective question of what constitutes a desirable balance between the national and non-national workforce and population, government discourse on labour force nationalization is becoming increasingly detached from reality. High fertility rates in the six states are decreasing faster than the Arab countries' average. On average fertility dropped from 4.7 children per woman in 1990 to 2.4 in 2009.[52] Among Emiratis, for example, later marriages and increasing education and employment among women are partly responsible for the rapid decline over the past decades.[53] In their official discourse, GCC governments continue to portray themselves as actively seeking to decrease the share of non-nationals. The large numbers of foreigners are portrayed as a threat to security, culture, identity, societal structure, and even foreign policy.[54] Paradoxically, however, sustained marginalization of nationals can also contribute to the ruling elites' survival in power by reducing political reform pressures, as political liberalization would inevitably entail a debate on the political rights of long-term foreign residents.[55]

2

THE EMERGING DOMESTIC
ENERGY INSECURITY

The massive influx of foreign workers to the Gulf monarchies in the 2000s was the result of increased economic activity enabled by high oil prices, coupled with a shortage of domestic labour. This followed a pattern familiar since the onset of the oil era. The difference this time was that the formerly invisible limits of national resource carrying capacities now became visible and tangible. The growth in population and industry outpaced the existing system of natural resource provision and infrastructure. Rising living standards and outdated pricing mechanisms for petrol, electricity and water, both 'symptoms' of the rentier state apparatus, contributed to the problem. Governments that were accustomed to thinking of energy security in terms of security of international oil demand, with low levels of domestic energy consumption relative to total oil and natural gas production, had not given much attention to issues of domestic energy security. Demand management was particularly neglected. The GCC monarchies were ill-prepared to deal with the fast economic growth and population explosion.

While consumers acutely felt the infrastructure crunch in the skyrocketing housing costs and swelling traffic jams, the looming energy shortage was left largely unnoticed by individual residents, since most of the few petrol and utility price increases were moderate. This, in turn, did little to curb the prolific natural resource consumption patterns. Alongside growing concerns over resource depletion in Oman, Bahrain, Dubai and the smaller UAE emirates, the late 2000s was the first time during the oil era that domestic energy security considerations rose onto government agendas and prompted a series of responses relating to alternative energy sources and energy efficiency and conservation.

The GCC energy sector itself is highly energy intensive and consumes a large share of energy use. In Qatar, for example, the oil and gas industry and its flaring practices accounted for 58% of total energy use in 2006, according to Qatar Petroleum.[1] Rentierism and the ruling bargains also contributed to the rising resource demand: low prices and consequent wasteful consumption patterns, as Hertog and Luciani have observed, are so ingrained that building codes and standards ignore energy and water use and efficiency.[2] This leads to enormous energy losses in the hot climate of the Gulf. Two further pairs of clashing elements push power demand levels up: first, the harsh local climate, with summer temperatures often reaching well over 40°C, clashes with a widespread habit of maintaining indoor temperatures at under 20°C; secondly, extreme water scarcity clashes with the resolve to achieve at least partial agricultural self-sufficiency and support large cities and segments of population with prolific water consumption patterns almost exclusively through desalination. Indeed, the Gulf monarchies have, because of all these factors, some of the highest per capita electricity consumption rates in the world: in 2009, Kuwait, Qatar and the UAE all ranked in the top-ten, according to the International Energy Agency, with average consumption rates nearly double the OECD average.[3]

The gas crisis

The Gulf resource crisis that emerged in the late 2000s was caused and defined by lack of natural gas in five of the six GCC states. The overall abundance of fossil energy resources and the consequent difficulty for governments to justify higher end-user prices were among the key factors. Paradoxically, the accelerating efforts to diversify the economy away from fossil fuel dependence also contributed to energy insecurity. Having been endowed with the comparative advantage of cheap energy, the GCC states invested in energy-intensive heavy industries as a diversification strategy. Until the recent decades, natural gas was often regarded as a useless by-product of oil production and often flared away. Its value only began to be recognized with increasing domestic power demand and advances in gas liquefaction technology, which allowed Abu Dhabi,

Oman and Qatar to export gas that had been regarded as economically 'stranded'.[4] At the same time, following stepped up diversification efforts in the 2000s, increasing quantities of natural gas continued to be consumed as feedstock by industries at a break-even price or a small profit.[5] Simultaneously, it was fed into the oil fields of the Arabian Peninsula for reinjection and enhanced oil recovery, to maintain pressure and production levels.[6]

Natural gas is the main and also the preferred electricity source in the Gulf monarchies because the opportunity cost of using it domestically is considerably lower than for oil. The UAE, Qatar and Bahrain currently use gas in nearly all power plants. Kuwait and Saudi Arabia, owing to limited gas availability in relation to demand, have resorted to using significant and increasing amounts of oil-based liquid fuels for electricity generation: up from 44% and 28% in 1980 to 71% and 55% in 2009, respectively (see table 6). Oman produces a fifth (18%) of its electricity from liquid fuel, while simultaneously exporting natural gas under long-term contracts to Asia.[7]

Most of the region's natural gas is associated gas, production of which is linked to OPEC's production quotas. Outside Qatar, new sources of non-associated gas have been difficult and expensive to find and develop, as recent discoveries of high-sulphur or tight gas attest.[8] Also, Qatar, the UAE and Oman have tied a considerable share of their gas production to long-term liquefied natural gas (LNG) export agreements to Asia and Europe. Even in the early 2010s, most retail electricity prices remained far below production and distribution costs, let alone international levels. As table 6 shows, according to data collected by El-Katiri, the price of a kilowatt hour (kWh) for the residential sector in the GCC states in 2007/8 was US¢0.3–2.2/kWh, while prices reported in EU countries were US¢15–34/kWh and in the United States US¢11/kWh. The situation in the industrial sector was similar: US¢0.1–2.2 compared to US¢7–24 in the EU and US¢6 in the US.[9]

As a result of the GCC states' failure to develop their own gas fields apace with the growing domestic demand (partly because of the discouraging effect of subsidized prices), supply of natural gas was increasingly unable to meet demand.[10] Industrial projects like a massive aluminium smelter in Al Ruwais in Abu Dhabi were put on hold. Power plant operators across the region were forced to switch to more expensive liquid fuel.[11] Two clear exceptions to this pattern were Qatar, which had already

Table 6: Natural gas production, consumption and consumer prices.[12]

	Natural gas production/ primary supply (Mtoe, 2009)	Natural gas net imports (Mtoe, 2009 [2010 est.])	Share of gas in electricity generation (2009)	Electricity prices for residential sector (US¢/kWh, 2008)	Electricity prices for industrial sector (US¢/kWh, 2008)
Bahrain	7.9/7.9	–	100%	0.32	1.70
Kuwait	9.4/10.1	0.7 [2.3]	28.8%	0.28	0.14
Oman	23.5/10.9	−12.7 [−2.3]	82.0%	1.04	1.65
Qatar	79.5/19.5	−60.0 [−86.2]	100%	0.88	0.77
Saudi Arabia	61.4/61.4	–	44.8%	0.54	1.30
United Arab Emirates	40.9/49.1	8.2 [7.9]	98.2%	2.18	2.18

in the 1990s begun developing its enormous natural gas reserves for export and economic diversification, and Saudi Arabia, which neither exports nor imports natural gas but burns crude oil for domestic consumption. Other monarchies felt forced to seek regional and even international sources of gas imports. In 2007, Oman and the UAE began importing gas from Qatar via the Dolphin pipeline (56 million cubic metres per day in 2009).[13] In 2009 and 2010, as Kuwait and Qatar could not agree on the price, the former ended up importing LNG for the summer months' demand peaks from outside the region.[14] Also, in 2010, Dubai started receiving LNG from Qatar. Bahrain, in turn, has explored importing of gas from Russia, and Iran and Sharjah are connected by a pipeline.[15] By 2012, however, in a reminder of the region's geopolitical complexity, several of the monarchies had been in bilateral talks with Iran on gas imports but none had been concluded successfully.

Challenges related to power generation capacity bring another layer to the problem. Owing to extremely hot summer day temperatures, often well over 40°C, electricity demand in this season can be up to 40% higher than in the winter season, which requires the countries to maintain a significant capacity that is unused outside peak consumption months.[16] Persistent outages affected several GCC states throughout the summer in the late 2000s. These were due to infrastructure deficiencies, lack of gas, and unwillingness to pay high prices for liquid fuels.[17] In 2009–10 the emirate of Sharjah, Bahrain, Kuwait and Saudi Arabia all experienced repeated summer power cuts. Notably, in the UAE, the inter-emirate level brings an additional layer of dependencies since Abu Dhabi provides subsidized fuel and electricity to the other emirates. While the UAE's four northernmost emirates were reported to suffer from chronic power shortages throughout the late 2000s, owing to insufficient supplies of cheap gas, the arrival of the Arab Spring quickly ended this plight, with Abu Dhabi announcing in early 2011 a massive electricity and water supply package to its poorer neighbours, which have important Emirati populations.[18]

In a longer historical perspective, domestic consumption of electricity in the GCC states has grown very rapidly throughout the past three decades, reflecting the fast socio-economic development. In proportional terms, the greatest increases between 1980 and 2000 took place in Oman (1014%), the UAE (533%) and Saudi Arabia (517%). During the same period, energy consumption in OECD Europe grew by only 57%. As is

Table 7: Power and energy demand growth in the Gulf monarchies.[19]

	Total electricity generation (GWh, 1980)	Total electricity generation (GWh, 2000)	Total electricity generation (GWh, 2009)	Average annual generation growth rate (CAGR,* 2000–09)	Energy self-sufficiency (2000/2009)
Bahrain	1,160	6,297	12,056	7.5%	2.5/1.9
Kuwait	9,023	32,323	53,216	5.7%	6.1/4.3
Oman	818	9,111	17,823	7.7%	7.4/4.5
Qatar	2,416	9,134	24,796	11.7%	5.6/5.9
Saudi Arabia	20,452	126,191	217,082	6.2%	4.7/3.4
United Arab Emirates	6,306	39,944	90,573	9.5%	4.6/2.8

* Compound annual growth rate.

shown in table 7, in the 2000s power generation in the six states grew by 65–171%, equal to an average annual growth of 6–12%. Average annual growth in OECD Europe in this period was merely 0.8%.[20]

While proportional growth of electricity generation and consumption in the two earlier decades was extraordinary and reflected the fast socio-economic transformation from the pre-oil to the oil era as well as population and economic growth, it was not as significant in absolute terms as in the 2000s. In Qatar (see table 7), between 2000 and 2009 absolute electricity generation growth in the country was 15,700 GWh. In the two previous decades (1980–2000), generation grew by only 6,700 GWh. In absolute terms Saudi Arabia, owing to its economic and population size, is set clearly apart from the five smaller states, with a total consumption several times greater, and hence challenges on a different scale.

A further way to look at the 2000s' changing domestic demand situation is energy self-sufficiency (measured by energy production/total primary energy supply). The Gulf monarchies are still among the most energy self-sufficient countries in the world, but despite increases in oil and gas production capacity and volumes, this ratio fell in all states except Qatar during the past decade. The opportunity cost of consuming oil and gas domestically becomes tangible when domestic energy consumption (total final consumption) is juxtaposed with export figures: by 2009, consumption in Bahrain, the UAE and Saudi Arabia had climbed to correspond to 54%, 34% and 26% of their energy exports, respectively. In the three other states the shares were 12–14%.[21]

Since the advent of hydrocarbon production, the total energy mix of all six states, including fuel consumption, has been dominated by natural gas and oil. In the case of electricity supply, until the late 2000s three options were perceived as realistic: domestic gas, imported gas, and domestic oil. Because of the economic and political obstacles relating to gas imports described above, domestic gas was the preferred choice, but in most cases there was not enough to cover demand. With international prices rising, using oil for domestic consumption created growing opportunity costs for governments. As a consequence, around 2006–07, most Gulf monarchies became vocal about wanting to replace or complement oil and gas domestically with alternative sources, such as nuclear energy and renewables. Around the same time, the environmental consequences of burning fossil fuels, already prominent on the international agenda, began to influence domestic choices, particularly as they coincided with government interests regarding oil. In other words, a higher carbon footprint became another justification for not consuming oil for domestic power generation. While there are as yet no viable alternatives to replace oil as a transport liquid fuel, alternative sources for power generation, principally nuclear, solar and even wind energy, became increasingly perceived by Gulf economists and governments alike as potential alternative sources for electricity production.[22]

The energy shortages also had some positive consequences, in the form of speeding up intra-GCC energy cooperation and introducing a new energy provision option: electricity trading. In 2009 the first phase of the GCC-wide power grid, which had been planned since the establishment of the council in 1981,[23] was completed and the national grids of Bahrain, Kuwait and Saudi Arabia were connected. The UAE joined in 2011 and Oman was expected to join in 2013. In 2010, the grid helped the connected countries to avoid power cuts, with up to 1,000 MW solicited by Kuwait in the summer of that year.[24] As a result of electricity exchanges, 5,000 MW, or more, of total savings in capacity expansion are estimated by 2030. The main purpose of the intra-GCC grid is to provide emergency supply, but it will also allow for regional energy trading, and perhaps a regional electricity market later on. Some have even envisaged an eventual connection of the GCC grid to Europe, or the proposed Desertec energy supergrid that would export renewable energy from the Middle East and North Africa (MENA) region to the EU.[25] The US$1.6bn Desertec project is also seen as a possible catalyst for expanding cooperation into other natural resources, including water, gas and transport. Feasibility studies for a GCC water grid with encouraging results were carried out in 2003–08, but never acted upon.[26]

Enter the new energy future?

For long, diversifying the domestic energy supply base, to increase system stability, was not considered a priority in the GCC. This only began changing in the late 2000s. Small-scale solar and wind energy and R&D developments in Saudi Arabia, Kuwait and Bahrain date back to the 1970s, but the knowledge gained was never transferred to the industrial sector. This was due to a lack of a 'socio-economic perspective' behind the early projects. In the 1980s and 1990s, two villages in Saudi Arabia were provided with electricity from 350 kW PV solar plants built jointly with US and German assistance. In 2011, signalling a revived interest in solar energy in the country, a 500 kW plant was inaugurated on the Red Sea island of Farasan.[27] In the smaller monarchies, renewable energy projects only appeared in the 2000s and deployment was similarly for long limited. One example is the 850 kW wind power project on the remote Sir Bani Yas wildlife reserve island of Abu Dhabi, commissioned by the ruler Sheikh Zayed bin Sultan Al Nahyan and built in 2003–04.[28] Larger-scale plans and projects only began appearing in local newspapers at the time of the late 2000s energy crunch. These included a 100 MW solar plant in Oman, a 100 MW solar desalination plant and a total of 3.5 GW (US$4bn) in solar capacity in Qatar, and a 'solar island' project in Ras al-Khaimah.[29] In what is becoming perceived as a major game-changer in the region, Saudi officials have suggested that the country would invest over US$100bn in solar and nuclear energy, each, over the next two decades. Pushed forward by the high and growing opportunity cost of burning oil domestically, Saudi solar plans in 2012 included the creation of a 5 GW capacity by 2020 and possibly up to 41 GW by 2032, with investments coming from public and private sources.[30]

By 2012, however, most projects elsewhere in the GCC had still not progressed from the planning stage, generally because of their lack of economic viability in the existing utility price regime. Small-scale solar projects had only begun appearing on Gulf cities' rooftops, with Abu Dhabi leading the way with 11 office buildings with a total capacity of 2.3 MW. Wind pilot projects had also been announced in Oman and Bahrain, and plans for a 20–30 MW wind farm were considered in Abu Dhabi.[31] Many renewable energy projects were discarded as a consequence of the economic crisis, including a solar panel production plant with an annual capacity of 130 MW in Dubai.[32] Furthermore, even though they were pioneering in nature, completed projects like Bahrain's World Trade Centre (2008)—the world's first skyscraper with integrated wind turbines—seemed to have few positive spillover effects on their surroundings, and were even rejected by them: Bahrain's WTC could not feed its excess electricity to the local grid, for lack of legislation.[33]

The only larger project in the renewables sector that had been completed by the end of 2011 was a 10 MW PV solar plant in Masdar City, owned by Abu Dhabi's Masdar Initiative. Masdar, established in 2006, was also working on a 100 MW concentrated solar power plant, with a plan to expand capacity to roughly 1.5 GW by 2020. In addition to solar plants, Masdar has invested abroad in renewables technologies and estab-lished a research institute, and is working with different types of carbon reduction projects. In Qatar, the Qatar Science and Technology Park is leading a push in the area of R&D into renewables and energy technol-ogies. Both Masdar and Qatar's technology park will be analysed in detail in the coming chapters.

While solar power has until recently been regarded throughout the GCC as too expensive, given the existing subsidy regimes, and hence as something for the longer term, nuclear energy emerged in the mid-2000s as an attractive medium-term option for domestic source diversification and for 'saving' precious oil for export. Up to around 2006, it was broadly regarded that the GCC states' fossil fuel abundance, their military alli-ances with the United States and its bases in the region would keep these from considering nuclear energy programmes as long as Iran refrained from acquiring a nuclear weapon.[34] But in December 2006, on Kuwait's proposal,[35] the six states announced a study on the feasibility of a col-lective civil nuclear programme. Because of the timing and the previous lack of interest in nuclear power, the move was perceived by most observ-ers as a strategic challenge to Iran. However, it soon became clear that economic and energy security motives were driving the GCC, in addi-tion to nuclear supplier states keen to sell their technologies to new mar-kets. All six GCC states justified their interest in terms of domestic energy security, with the need to cut greenhouse gas emissions as an additional argument.[36] Multiple assurances of the civilian and peaceful nature of the programme followed, including pledges by both the UAE and later Kuwait of US$10m to one of the international nuclear fuel bank proposals.[37]

The GCC joint study, conducted by the International Atomic Energy Agency (IAEA), was completed in November 2007. The collective plan, however, has not advanced since then. Explanations for this include mutual distrust among the group's member states—including the fear of Saudi hegemony—, the security concerns relating to sharing a nuclear programme and exacerbated by this distrust, and the inertia of a multi-lateral process.[38] Notably, not a single joint nuclear energy programme exists in the world to date. The small grid size in countries like Bahrain and Qatar has also reduced the attractiveness of nuclear energy for these individual countries, but with the GCC grid in place, this might change in the near future.[39] Clearly, one of the reasons why the GCC project has practically withered was the UAE's decision to proceed unilaterally. As of 2012, neighbouring Saudi Arabia's ambitious plans to build a massive generation capacity by the 2030s were still on paper, and Abu Dhabi was the only Gulf monarchy pressing ahead with nuclear power, pursuing plans to build four 1.4 GW reactors by 2020. Abu Dhabi's nuclear energy programme will be analysed in chapter 5.

Although coal could be an affordable option—and a supply-wise safe one—for some of the monarchies the rise of climate change on the international energy agenda has made it less attractive. Abu Dhabi, for example, has discarded coal completely as environmentally 'detrimental'.[40] In 2008–09, Ajman and Ras al-Khaimah each made announcements on the commissioning of a 1 GW coal plant, but presumably promises from Abu Dhabi in 2010–11 to provide electricity to the northern emirates have, at least partially, delayed the projects.[41] Oman also announced (and later discarded) plans to build a 1 GW coal plant. And in 2011, Dubai declared that 'clean coal' would be a part of its energy diversification strategy to 2030.[42]

Carbon capture and storage (CCS), a series of technologies used for capturing carbon dioxide (CO_2) and injecting it into storage spaces like ageing oil reservoirs, has had receptive audiences in the Gulf. It is hoped that employing CO_2 instead of gas in enhanced oil recovery would allow producers to decarbonize oil production. CCS is therefore commonly seen as a way to extend the era of fossil fuels. CCS technologies are still largely at the pilot stage, however, and uncertainties regarding their economic viability and safety for large-scale use prevail. The Gulf OPEC monarchies have nevertheless been active in seeking to include them in a post-2012 UN climate treaty so as to ensure financial support from developed countries for their costly implementation, as will be discussed later.

In the past, Gulf governments largely ignored demand side management, and emphasis was put on guaranteeing domestic supply at any cost. In 2009, according to the International Energy Agency (IEA), Saudi Arabia spent roughly 10%, the UAE 5%, and Qatar 3% of its GDP on fossil-fuel consumption subsidies.[43] As noted by Hertog and Luciani, the logic behind the high subsidies for fuels is political, not economic. The domestic sector is the largest consumer of electricity in the Gulf monarchies. Lowering subsidies is, however, still a taboo owing to the existing social contracts, whereas the available alternatives for providing incentives for behavioural change, such as awareness-raising, are not nearly as effective.[44] In the residential sector, air conditioning consumes significant amounts of energy: according to some estimates it absorbs upto 70% of total electricity consumption of buildings.[45] Increased attention to energy efficiency of buildings and conservation at the individual level could bring financial benefits to consumers and governments, as this would ease the pressure to increase power production capacity, and even prices.

Savings could be achieved in the transport sector as well: in Qatar, for example, there were 470 vehicles per 1,000 inhabitants in 2007, which was close to the OECD average of 490.[46] This is significant given that a year later Qatar's statistical authority placed the proportion of manual labourers, who generally cannot afford a car, at 57% of the population.[47] For higher income groups, having fewer cars per family, using public transport, or raising room temperatures, would not require major sacrifices or economic investment, but a change of mindset. This is however extremely difficult, as the natural resource subsidies that extend beyond the nationals effectively maintain a population-wide rentier mentality, reflected in energy and water use patterns. Also, the high social stratification and prejudice towards lower income groups mean that public transport is often looked down upon. Many shun the idea of riding in the same vehicle with a construction worker; others simply prefer the comfort of a private car, running on cheap petrol. In addition, the expatriate populations, which in the cases of Qatar and the UAE comprise over 80% of the population, have a fast turnover rate: people 'passing by' are not likely to develop a sense of belonging in their host countries. Concern for the local environment is therefore not a priority for most expatriates.

Early signs of the unsustainability of the status quo, particularly in the weak rentier monarchies, have begun to appear in the 2010s. For example, the need for a gradual weakening of the subsidy regime is becoming more openly recognized by the governments. In late 2010, Bahrain's Oil Minister called for a rethink of the existing fuel prices subsidies. The country has also announced a 50% price hike for industrial users of natural gas in 2012.[48] However, at the same time, the regional uprisings and revolutions that began in 2011 strongly reduced the GCC governments' willingness to cut consumer subsidies in the coming years.

'Green' diversification strategies

In their effort to 'grow beyond oil', not only did the Gulf monarchies expand their 'old', fossil fuel-dependent economic sectors (energy and petrochemical industries), there was also a conscious, government-led push to accelerate economic diversification to 'new', non-oil sectors.[49] An emphasis on energy-intensive sectors like petrochemicals, heavy industries, and even real estate, further pushed domestic energy demand up. More interestingly, however, the GCC governments had their finger on the pulse of global trends in energy, building, lifestyles and values, and drew from these for their fast-evolving diversification agendas. Themes like low-carbon energy and renewables technology, sustainable building and eco-tourism, and knowledge and technology transfer in the areas of alternative energy and the environment emerged on each monarchy's policy agenda at differing speed, intensity and breadth.

The most active ones were Abu Dhabi and Dubai of the UAE, Qatar, and Saudi Arabia. Endowed with extravagant external rent surpluses and high government autonomy, these four devised some of the most innovative attempts at 'green diversification'. The first and most ambitious of all was Abu Dhabi with its multi-billion-dollar alternative energy initiative Masdar, established in 2006, that incorporated elements of technology transfer, research and development, green living, and alternative energy production. Abu Dhabi's increasingly ambitious government also made a controversial decision in the field of energy source diversification when it announced in 2008 the establishment of a civilian nuclear programme. Moreover, in 2009, riding the wave of positive attention showered upon Masdar, Abu Dhabi secured the headquarters of the International Renewable Energy Agency. In the area of green building, Dubai's government, highly image conscious and seeking to fight off a reputation as a 'haven of ecocide', came up with an even more ambitious plan: in 2007, the ruler Sheikh Mohammed announced a green building code to be enforced in all new buildings starting from 2008. The code has yet to be enforced, however, as Dubai, ironically the most diversified and dynamic monarchy in the region, took a hard hit from the global economic crisis.

Saudi Arabia devised a set of green diversification plans too, but because of inertia typical of larger states—plus a far more complex ruling bargain and the need to support several times more nationals than those in the smaller monarchies—real movement was only visible in terms of new research priorities and centres. These included initiatives and building features linked to the new King Abdullah University of Science and Technology, opened in 2009, as well as research in two Riyadh-based research centres. Despite its overwhelming natural gas abundance, Qatar did not stand aside and ignore the opportunities of the emerging global climate regime. By 2007, Qatar Petroleum had set up the region's first and largest greenhouse gas emission reduction project at the al-Shaheen oilfield, entitling it to monetize its emission savings and sell them as carbon credits under the UN Clean Development Mechanism. Similarly to Abu Dhabi and Saudi Arabia, economic diversification efforts in the green and low-carbon sectors in Qatar first materialized in the form of research initiatives, with the Qatar Science and Technology Park at the forefront. Some progress was also made in the field of sustainable building. Also, green promises made as part of Qatar's successful bid for the 2022 FIFA football World Cup in 2010 gave a time frame for a number of sustainable urbanistic projects, including metro (already in place in Dubai) and rail networks and enough installed renewable energy capacity and other offsetting actions to create a carbon neutral event.[50]

External pressures: climate change and international demand

The themes of alternative energy, low-carbon and green did not emerge from the GCC states' domestic context alone, but were invented as the result of, and in continuing dialogue with, emerging external pressures, opportunities and trends. The high level of globalization and openness to modernity and global influences, in the smaller Gulf monarchies in particular, directly exposed them in the late 2000s to the issue of climate change. In 2007–09, with the publication of the Fourth Assessment Report of the Intergovernmental Panel on Climate Change (IPCC) and the momentous United Nations climate conferences in Bali and Copenhagen, climate change established itself more firmly than ever on the global agenda. As affirmed by the IPCC report, judging by rising global temperatures and melting of snow and ice, warming of the climate system is 'unequivocal' and can be firmly attributed to human activity. If greenhouse gas concentrations in the atmosphere were to rise to a level that pushed global temperatures up by over $2°C$, this would be likely to have catastrophic consequences, including sea-level rise, changing temperature and rainfall patterns, and increased and intensifying extreme weather events. This, in turn, would lead to increased water stress, ecosystem damage, challenges to food production, more harmful floods and storms, health risks, and major pressures for adaptation to rising sea levels for coastal settlements and infrastructure.[51]

With the rise of climate change, fossil fuels which had been the Gulf monarchies' black pearls, a gift from God, and were seen as a source of wealth for future generations, officially became a global problem. This reality was hard for the region's people and governments to assimilate. In a region rife with conspiracy theories, it was easy to think or claim that climate change was yet another plot against the development of the Arab Middle East.[52] Climate change was simply something that did not fit in the 'business as usual' plans of Gulf monarchies' governments regarding their political economies and social contracts. Climate change disturbingly brought the topic of post-oil era closer; it increased the sense of urgency to diversify Gulf economies away from oil.

Another form of denial that emerged was the argument that Middle Eastern countries had nothing to do with the problem, and even if they did, they should not be held accountable as mere producers of fossil fuels, and would have to be compensated not only for any climate change adaptation-related costs, but also for revenue losses related to climate change mitigation. Indeed, the oil producers' responsibility for the past environmental externalities of their accumulated wealth was and is a complex moral issue, as is the industrialized countries' responsibility for historic emissions at a time when the consequences of emitting carbon dioxide were unknown. However, it was clear by the late 2000s that the oil and gas exporting Gulf monarchies would have to face the environmental costs of their chief exports. Pressure would come in two main forms: through changing international energy demand patterns and policies, and through the politics of the international climate regime.

Different conceptions of energy security

For the Gulf oil and gas exporters, the ongoing transformation of the current energy paradigm entails increasing uncertainty over the future demand for their exports. The key energy security concern for these states therefore is, besides the duration of domestic reserves and their availability for domestic use and export, the stability of international demand and prices of fossil fuels. Regardless of increasing global energy demand, Middle Eastern oil exporters have not managed to regain the leverage they exerted on prices in the 1970s, because of increased supply from non-OPEC member states and expansion of alternative resources.[53] As discussed above, in many Middle Eastern states export capacity is affected by diminishing fossil fuel reserves and a simultaneous trend of growing domestic energy consumption, prompted by population and economic growth and social development. Some oilfields in the Gulf area are mature, and non-OPEC exporters Bahrain and Oman, as well as Dubai, are coping with falling oil production.[54]

Energy importers, in turn, stress security of energy supply. Since the first international oil shocks in the 1970s, sustainable and secure energy supplies have figured as a central concern in international politics. From around 2003, increasing oil prices led again to a remarkable rise in the global significance of energy security. Simultaneously, countries became aware of the urgent need to integrate energy policies with climate change mitigation.[55] As a result, climate change and energy security became almost inseparably entangled with each other—and with foreign policy. Climate change emerged as an internationally important issue already in the late 1980s,[56] but its rise on the international agenda was significantly accelerated by the four reports of the IPCC in 1990–2007. In its fourth assessment report the IPCC declared that the warming of the climate system is caused by the emission of greenhouse gases by humans, with more than a 90% probability.[57]

From this period climate change, both as a phenomenon and as a politicized issue of international relations, began having significant foreign policy implications globally. This enmeshment also pushed some key aspects to the core of both international energy politics and the concept of energy security, such as diversification of energy sources, energy efficiency, increasing the share of sustainable and low-carbon energy sources, and reducing carbon dioxide emissions. Moreover, it became generally accepted that there was no 'silver bullet' solution, and that bringing global greenhouse gas emissions to safe levels (a rise of less than 2°C from pre-industrial levels) would require a broad range of actions, including increasing the share of alternatives to fossil fuels, decreasing the carbon content of the fossil fuels still used (with the help of technologies), improving energy efficiency (of the built environment, in industry, and of vehicles), and setting price-driven instruments (taxation or cap-and-trade) to change consumption patterns.[58]

The unknowns of future demand and supply

The price of oil is still the most significant external instability factor for the six GCC economies. The logic of oil prices is different from that of natural gas, which is still not traded on an international market and is mostly sold through long term contracts with set prices.[59] The price of oil has generally followed a boom-bust cycle and the ability of OPEC to control pricing is weak owing to the self-seeking behaviour of the member states, erosion of spare capacity, and speculation, among other things.[60] Sudden price falls can prove disastrous for economic and political stability when producer states plan their budgets on the basis of higher price levels: most budgets of the Gulf monarchies for 2009 were not prepared for prices below US$50/barrel. Such actions, however, only work in the short term. Moreover, periods of lower prices tend to have a negative impact on the industry because of the scaling down of investments, cost-cutting strategies, reduction in R&D spending and lack of ability to attract students in the field.[61] Signs of this were visible in late 2008 when oil prices plunged and Saudi Arabia, Kuwait and the UAE were forced to revise important oil production capacity expansion projects.[62]

In the medium and long term, global supply and demand and the price of oil will be driven by several factors, all of which involve important uncertainties. These factors include growth in the major Asian economies, the national energy efficiency and security policies of consumer states, climate change abatement policies globally, and the development of alternative energy technologies and infrastructure and of clean fossil fuel technologies. The physical and social impacts of climate change may also speed up the global transition to a non-oil energy paradigm.

As was pointed out earlier, the economies of the Gulf monarchies still remain highly vulnerable to swings in global demand for oil and natural gas. While Saudi Arabia, Kuwait, Abu Dhabi and even Qatar (relatively to its population size) still report considerable remaining proven oil reserves, the IEA, for example, has implied that official data provided by the governments may be exaggerated to allow for larger production quotas under OPEC. The sudden jump in official oil reserve figures of Kuwait in 1984 and the UAE in 1986 and the consequent balancing within a very small margin (93–102bn barrels (bbl)) indeed leave room for speculation, as do persistent claims that reserves have been significantly over-estimated.[63] If the alleged data inflation is significant, the consequences not only for global energy security, but for the states' economic stability in the future could be important. Notwithstanding the relatively abundant natural gas reserves of these same four states, the only one with export expansion potential is Qatar, with 25 trillion cubic metres, or 14%

of global proven reserves, in 2010.[64] Even here, a self-imposed multi-year moratorium on new production has elevated uncertainty and speculation over the total recoverability of Qatar's reserves. The other states have little potential to expand production in the medium term, as they are already using most of their gas domestically, and their gas production is largely tied to that of oil. Oman and Abu Dhabi still export gas, but owing to their increased domestic demand Qatar supplies both through the Dolphin pipeline.

For Saudi Arabia, Kuwait and Abu Dhabi, therefore, the key external energy security concern is how long and at what price the world will keep consuming oil. For Qatar this question applies to natural gas. The medium- and long-term estimates of global energy consumption—including, among the most quoted, the energy outlooks of the International Energy Agency (IEA) and OPEC—are extremely important tools as they help determine the need for investment in additional production capability on the one hand, and the urgency of economic diversification efforts in countries heavily dependent on energy export revenues, on the other. The IEA *World Energy Outlook 2010* central scenario for 2008–35 estimates global primary energy demand as likely to grow by 36%, with most coming from non-OECD countries. In this scenario, oil remains the dominant fuel (albeit with a diminishing share due to higher prices and fuel efficiency) and its demand is projected to grow by 18% during the period; while in a scenario that aims at limiting global temperature increase to 2°C, oil demand is expected to peak by 2020. Under the central scenario, OPEC output keeps rising and its share of global production will increase from 41% to 52%, and substantial new gross capacity will be needed to offset decline in production. Owing to climate change-related policies and measures, however, oil demand will remain 'highly sensitive', while demand for natural gas is expected to increase by 44%.[65]

OPEC's *World Oil Outlook 2010* paints a slightly more positive picture for oil, with total global energy demand rising by over 40% by 2030. Oil demand in the reference scenario grows by 25% by 2035, and oil remains as the lead fuel (over 30%), albeit with a slightly falling share. Natural gas use is also expected to grow fast in this scenario, partly because of the growing importance of shale gas in the US and elsewhere. The OPEC *Outlook* also observes that oil demand projections have been constantly revised downwards in the 2000s owing to climate change policies and later the economic downturn.[66] Both IEA and OPEC scenarios project high growth for coal but only modest growth for renewable and nuclear energy. Oil, it seems, will be king, at least for the next quarter of a century, and gas for even longer, but our ability to predict the future should never be overestimated.

In the short and medium term, price will be the main source of uncertainties, along with the potential for new reserve additions. The numerous factors that influence international oil prices, including supply and demand, energy policies, and the state of the international economy, make reliable price projections close to impossible. The IEA's 2010 projection of US\$113 per barrel for 2035[67] is one among many estimates. The GCC OPEC members have traditionally been interested in oil prices that are not so low as to harm exporters' supply and revenue prospects, but not so high as to endanger economic growth in importing states.[68] High oil prices also increase the economic feasibility of competing energy sources and investment in energy efficiency and savings. According to Mitchell, in 2008, this limit was US\$60.[69] Despite the public expenditure increases that have followed the onset of the Arab Spring (increasing the break-even price for most GCC states), the Gulf OPEC states have in the past years held US\$70–80 as a 'good price'.[70]

Energy export diversification is an effective way for producing states to buffer the negative impacts of price downturns on the economy. Qatar has greatly benefited from its extremely ambitious LNG export programme initiated in the late 1990s. Other monarchies have been less successful or fortunate. By 2011 only two other monarchies, Abu Dhabi and Saudi Arabia, had realized the need to engage with, and even seek a stronger position in, the low-carbon future of global energy. Qatar's rise as the world's largest LNG exporter, and Abu Dhabi's quest to become the region's 'alternative energy leader' are perhaps the best examples in this sense, as will be discussed in the coming chapters. Alongside diversification into areas where the GCC OPEC monarchies have a comparative advantage due to their plentiful domestic energy resources (within the limits of the domestic energy security situation), diversifying into other energy export products, particularly solar power, is indeed probably the best chance the local monarchies have in the long term of extending their external rent, and consequently their domestic rentier bargains, beyond oil.

Despite the obvious benefits from Abu Dhabi's and Qatar's non-oil energy strategies, there are some important unsustainabilities that will continue to hinder development efforts, in particular with regard to find-ing a balance between natural resource use and environmental sustain-ability. In the case of Abu Dhabi, two factors increase the uncertainty regarding continuity of new, promising policies and projects: the high concentration of natural sustainability efforts under one patron, and the vulnerability of the nascent renewable energy and sustainability indus-try to economic and political shocks. Despite Abu Dhabi's increasing assertiveness and independent manoeuvring, the UAE's federal structure also complicates decision-making. On account of diverging needs and interests among the emirates, the federal structure considerably widens the spectrum of challenges that the leading emirate is expected to tackle so as to maintain domestic welfare and stability. The growth of Qatar's LNG economy, in turn, has taken a heavy toll on natural sustainability within the country's borders. The energy industry damages Qatar's marine environment and generates enormous amounts of greenhouse gas emis-sions and air pollution. The abundance of natural gas also serves as a pow-erful disincentive for the government to cut fuel and utility subsidies, leading to superfluous consumption and sheer waste.

3

FACING UNSUSTAINABILITY

THE CHALLENGES OF CLIMATE CHANGE

Politics of the international climate regime[1]

If one issue only should be named as prompting the rise of climate change on Gulf monarchies' policy agendas, this would be greenhouse gas emissions. The economic dependence on the source of these emissions has been central in shaping the GCC states' policy positions vis-à-vis the international climate regime, while the high per capita levels have attracted new and uncomfortable international attention. The international climate regime is defined in this study as negotiations taking place under and around the United Nations Framework Convention on Climate Change (UNFCCC), including the Kyoto Protocol, and other institutions and fora that seek to multilaterally prevent dangerous climate change. The limit of dangerous climate change is defined as an average global warming of over 2°C from the pre-industrial levels. In the UNFCCC, countries are divided into historico-geographical groups. The main division lies between the developed and developing countries (according to a classification agreed upon in 1992), with questions of responsibility and equity at its core. The two main groups under the UNFCCC are the Annex I countries, or the industrialized countries and the transition economies that have committed to return their greenhouse gas emissions to 1990 levels and below; and the Non-Annex I countries, or those classified as developing countries that have ratified or acceded to the convention but have lesser commitments towards it. Secondary divisions are constituted by regional and political groups. While the developing countries' umbrella group is the G77+China group of roughly 130 members, there are several other formal and informal negotiating groups under it. As of 2010, the main developing country subgroups were: BASIC (Brazil, South Africa, India and China), AOSIS (Alliance of Small Island States), LDCs (Least Developed Countries), the Africa group, OPEC (led permanently by Saudi Arabia), and the newly-emerged ALBA (the Bolivarian Alliance, led by Venezuela).

Although the industrialized countries bear the historical responsibility for climate change, the 'future responsibility' (albeit a very contentious issue) will lie on the developing countries, particularly large emerging economies, such as China and India, where emissions are growing fast.[2] Simultaneously, however, the principle of the developing countries' right to sustainable development is enshrined in the United Nations' values.[3] Thirdly, of importance for the fossil fuel exporters is the question of responsibility for the emissions of their oil (and natural gas) exports: GCC OPEC states have been advocating the polluter (or beneficiary) pays principle, despite having benefited for decades from revenues from these products that have negative environmental externalities.

For the six Gulf monarchies, the most significant reference groups have been the Gulf Cooperation Council, OPEC and the G77+China. OPEC is a generally tightly disciplined group of structurally very different states that has traditionally held sway in the G77+China group on certain specific issues. However, of the GCC states, only Saudi Arabia, Kuwait, Qatar and the UAE partake in the OPEC group's coordination. Additionally, the member states of the Organization of Arab Petroleum Exporting Countries[4] coordinate their positions, with only Oman not belonging in this group. The functioning and relevance of the League of Arab States as an interest aggregate, despite an all-inclusive membership and increasing attempts at coordination, has for the most part been invisible, until the past two years or so.

Owing to the consensus-based negotiating system, small groups of states can in theory prevent an agreement from emerging. This was most visible in the infamous Copenhagen round of negotiations in December 2009 where Sudan, a few ALBA states and Tuvalu prevented a pre-negotiated accord from being endorsed (while the conference merely 'took note' of it). Arguably, throughout the history of multilateral climate negotiations certain OPEC member states have held positions that, if heeded to would effectively block ambitious[5] international climate change mitigation and hinder adaptation in developing countries. As will be demonstrated in chapter 8, Saudi Arabia in particular, but also the other GCC OPEC states, have deliberately aimed at slowing down the negotiations so as to secure the role of oil in the global energy economy. The OPEC states have made a particular impact on progress on the adaptation agenda by insisting that their demands regarding the negative impacts of international mitigation should be advanced at a similar pace to other issues on the agenda. Although Saudi Arabia is among the global top-20 emitters, the GCC states' total contribution is still small, estimated at 2.4% in 2007. Limiting the global temperature rise to 2°C will therefore depend above all on the major emitters of greenhouse gases, namely the United States (representing 20% of global CO_2 emissions in 2006) and China (22% in 2006), but also India, Russia, Japan and the EU.[6] Consequently, in the big picture of emission cuts, the GCC states remain small players.

By early 2012, many questions relating to ambitious international action to prevent dangerous climate change were still open and lacked a robust political solution. These included the application and meaning of the principles of historic responsibility and equity; the division of labour among the developed and developing states in determining the corresponding commitments; an efficient framework for global mitigation, adaptation, financing and technology transfer; and the actual concrete roadmap for the implementation of the post-2012 regime of international climate politics. Based on the principle of common but differentiated responsibilities (and respective capabilities) enshrined in the UNFCCC convention text (Art. 3.1)—according to which developed countries 'should take the lead in combating climate change and the adverse effects thereof'—the first commitment period of the Kyoto Protocol (2008–2012) carries greenhouse gas emission reduction commitments for only the most developed, or Annex I countries. The other, non-Annex I parties are only obliged to monitor and report their emissions. Currently, a new agreement is expected to be agreed upon by 2015 and take effect in 2020. It is envisaged to bring new commitments to all countries, with those by the developing countries being supported with financing and technology from the developed countries. While most pressure has been on developing economies that are major greenhouse gas emitters—China and India—the question of additional commitments for states with a high GDP per capita (but currently classified as developing states under the climate convention) was debated during the negotiations on the post-2012 treaty.[7] However, because of the small total emissions of most of these countries and low political ambition among key players—including the United States' inability, due to its domestic political situation, to sign into any internationally binding agreement to cut greenhouse gas emissions—in 2012, this issue still remained in the margins of negotiations.

Despite an apparent unity among OPEC and GCC countries in the international climate negotiations, these represent different degrees of strength of the rentier state and often different leadership interests and perceptions, which in turn have led to different positions in the past; pre-Ahmadinejad Iran, for example, held a more cooperative attitude, while Saudi Arabia has generally been against all attempts to move away from fossil fuels.[8] The potential negative impacts of international mitigation measures on the demand and price for oil are not as relevant for the two non-OPEC Gulf states, Bahrain and Oman, as for their wealthier neighbours whose fossil fuel reserves are expected to last for several decades, even centuries, at current production rates. Owing to their lower adaptation capacity, Bahrain and Oman could arguably benefit more from the advancement of the broader agenda than from ensuring that there is a mechanism for compensation for any future revenue losses. For oil-rich Saudi Arabia, Abu Dhabi and Kuwait, and gas-rich Qatar, the status of oil and natural gas in the global energy economy remains crucial. Because of these structural realities, dissimilar domestic responses and international policy positions were seemingly bound to appear at the turn of the 2010s, as climate change climbed towards the core of international relations—and indeed they did. Surprisingly, however, the first state to step forward with a more balanced agenda—the United Arab Emirates, led by Abu Dhabi—was anything but oil-scarce and poor.

Degrees of vulnerability[9]

Since the late 2000s, climate change has been a major driver in the international energy economy as both developed and developing states feel the pressure to mitigate its negative impact by shifting towards low-carbon economies. Climate change is also increasingly recognized globally as a challenge multiplier that can produce new sources of threats and instability for states. In the largest Gulf oil exporting monarchies, however, the global energy transition has in the past couple of decades been perceived as a more urgent and tangible source of instability than the potential threats of climate change itself. This is because the imminent global shift away from fossil fuels pushes the oil revenue-dependent rentier states to hasten domestic economic diversification towards either alternative sources of external rent or new economic, and consequently political, models.

The majority of international climate scientists agree that climate change is expected to create new kinds of challenges and threats to the stability of states, such as coastal flooding and food and water insecurity.[10] Although it is generally considered that climate change alone cannot cause conflicts, since around 2007 it has been described in Western security literature as a 'threat multiplier' that has the potential to complicate pre-existing problems and instabilities, thereby inducing 'multiple chronic conditions'.[11] As an important indicator of an elevated importance in international relations, climate change was discussed in the UN Security Council, as a result of a British initiative, in April 2007.[12] In June 2009, the UN General Assembly passed by consensus a resolution recognizing that adverse impacts of climate change could have possible security implications. This time, the initiative came from small Pacific island states that perceive climate change as literally an existential threat.[13] Because of the transboundary nature of the problem, small emitters of greenhouse gases with little historical emissions, including countries in the Middle East, will also suffer from the potential negative consequences of climate change—possibly even more than the major emitters (China, US, EU, Russia, India and Japan). Indeed, the Middle East is considered to be one of the regions in the world most vulnerable to the negative impacts of climate change.[14] The Gulf monarchies' physical vulnerability stems primarily from their already hot, desert-like climate and their concentration of population and infrastructure in low-lying coastal areas.

Vulnerability in the Middle East is, however, uneven. While the physical impacts may in many cases be relatively homogeneous, some states are in a better position than others to weather the consequences. In the short and medium term, most GCC states are arguably better placed in terms of strength of the state and economic capacity than most of their regional Arab neighbours. The abundance of hydrocarbons translates into assets, such as the availability of cheap energy and large financial resources.

These, in turn, can be converted into enhanced adaptation capacity ahead of future climate change-induced challenges, including extreme climatic conditions and increasing water and food insecurity. As long as energy resources and external rent are maintained, the Gulf monarchies will at least in theory maintain their capability to sustain their energy-intensive modern lifestyles (with air conditioning and seawater desalination); secure food supply through subsidized local farming and foreign farmland purchases; continue the opulent land manipulation projects; and, generally, keep up a strong state capacity through rent distribution, despite rising temperatures and sea levels.

In addition to these climate change-related instability factors, the internal stability of the oil exporting Gulf monarchies in the coming decades will arguably depend largely on both the international demand for oil and their ability to sustainably diversify their economies away from oil revenue dependence. Evidently, as long as the GCC states rely on a political economy that is based on rentierism, their governments' immediate interest is the maintenance of external revenues. It is therefore only logical that political elites in GCC states with vast remaining oil and natural gas resources may perceive global mitigation as an indirect threat to their countries' economic growth and political stability. Indeed, potential economic losses induced by international climate change mitigation have had the most weight in the considerations of governments in oil export-revenue dependent Gulf monarchies regarding the different types of possible negative consequences of climate change. Although there is not yet concrete evidence of such losses, these have unquestionably constituted the main negotiating issue for the group of OPEC countries in international climate negotiations, as will be discussed in more detail in chapter 8.

Natural scarcity and climate change

As Mathews has pointed out, population growth has been identified as the source of most environment-related trends, and fast growth rates can overwhelm any government.[15] In the late 2000s, the Gulf monarchies' demographic and economic circumstances, together with the rentier system, created a situation in which not only energy resources became scarcer, but other resource-related problems began surfacing. As a consequence of growing domestic demand for desalinated water, increasingly scarce renewable water resources, and extremely high dependence on food imports, water and food security emerged among many governments' top priorities. Environmental degradation and pollution also became more visible than ever, owing to rapid, unplanned growth. These domestic developments interplayed with the simultaneous transformation of the international agenda relating to energy security and climate change, arousing in some cases passive or defensive responses, in others proactive responses.

A major gap in both historical data (time series on past climate and weather patterns and groundwater aquifers) and region-specific future projections complicates estimate of the physical and social impacts of climate change on the Middle East.[16] The MENA region is, however, considered to be among the most vulnerable to climate change.[17] As a consequence of climate change, average temperatures in the region could rise from current levels by 2.0–3.7°C by the 2070s, according to the IPCC and the UK Met Office. A model study by the UAE predicts temperature increases of 2.1–2.8°C by 2050 and 4.1–5.3°C by 2100, and a drier climate. Precipitation is generally projected to decrease, albeit with large spatial variability. An increase in extreme temperatures and incidences of extreme weather events are also considered possible. Expected consequences are, among other things, drought, decreasing availability of water and dwindling agricultural production. Sea-level rise is the third major physical consequence of climate change that could cause significant inundation of coastal areas.[18] The IPCC's Fourth Assessment Report of 2007 projects sea-level rises of 0.2–0.6 metres by the end of the century, or potentially much more, should ice cap melting accelerate.[19] The MENA agriculture and coastal areas are described by the World Bank as vulnerable to both temperature increases and sea-level rise.[20] Other vulnerabilities listed by a UAE report to the UNFCCC are the sensitive dryland ecosystems and public health.[21] Moreover, traditional environmental problems, including desertification, marine, coastal and air pollution, construction and demolition debris, water quality issues and the consequences of military conflicts (particularly in the case of Kuwait), already plague the GCC states.[22]

Together with the existing stress factors, including the unsustainable use of natural resources, population growth, and the region's history of conflict, climate change could precipitate a number of social and economic problems, including increased electricity and water demand, leading to relative resource scarcity; inter- and intra-state tensions over natural resources; declining returns in agriculture, leading to internal migration; loss of coastal areas to sea-level rise and seawater intrusion, leading to significant economic losses and migration; and a host of related consequences, such as poverty, unemployment, social instability and radicalization.[23] In addition to climate change-induced losses, negative economic impacts for Middle Eastern states could also be caused by international mitigation.

Linked to rising temperatures, the most important potential negative consequences of climate change for the Gulf monarchies are resource scarcity and insecurity in the areas of energy, water and food. Domestic energy demand can be expected to grow in all five states under a business-as-usual scenario, but climate change-induced higher temperatures would further increase the need for air conditioning and water, and consequently demand for electricity and desalination. However, compared with water and food security, energy security is not as critical an issue for most Gulf monarchies, since they can always turn to oil as a last resort—with Bahrain and Oman as the obvious exceptions.

All Gulf monarchies except Oman already suffer from 'absolute water scarcity'.[24] Most of the small Gulf monarchies rely almost completely on desalination for drinking water[25] and have emergency reserves of only two to five days.[26] Even so, they exhibit the world's highest water consumption rates, of 300–750 litres daily per person, according to some estimates, compared with 150–300 in the OECD. The agricultural sector consumes most of the total water used in the five states. Owing to consumption rates that are higher than recharge, groundwater reserves are quickly depleting and their salinity has increased.[27] It is estimated that Abu Dhabi, for example, will run out of directly exploitable groundwater resources in 20 years if current consumption patterns prevail and if more sustainable and efficient water management policies are not put into practice.[28]

Although the Gulf monarchies are relatively well adapted to their structural food scarcity, unsustainable food self-sufficiency policies, initiated in the 1970s, are still maintained in many sectors. In addition to high water use, agricultural production is highly subsidized. However, its contribution to GDP is generally very marginal[29] and the five smaller states are still completely or highly dependent on imports of basic food articles, including sugar, rice, wheat and flour (99–100%), meat (55–80%) and vegetables (27–81%). Saudi Arabia's dependence is currently lower, but because of rising water scarcity, the government has announced it will phase out wheat, soya bean and fodder production subsidies, the first by as soon as 2016.[30] While the impacts of climate change on water demand in the Gulf are uncertain,[31] increasing water shortages and salinisation of coastal aquifers caused by current practices and consumption patterns alone could destroy the last hopes of any significant level of national food self-sufficiency. Economically, high food import dependence can also become a heavy burden as the world's population rises and climate change threatens agricultural yields. Saudi Arabia bears the heaviest price, reflecting the size of its population: the Economist Intelligence Unit has estimated that the country's food import bill will double from US$16.7bn in 2008 to over US$35bn in 2020. For the five smaller states, the total sum is forecast to grow from US$10.3bn to around US$23bn in the same period.[32] Therefore, as a reaction to declining water reserves, growing populations, rising global food prices, and scary scenarios of future availability, many Gulf monarchies have, particularly since 2007–08, sought food security outside their borders by exploring purchasing, leasing and investing in farmlands in Asia and Africa.[33] These agricultural policies are controversial, as they could undermine food security in the producing countries and lead to tensions between the producing and importing country. Also, by renting land in authoritarian countries, through lease revenues, Gulf monarchies might indirectly foster the emergence of new rentier sectors in the host countries' economies.

Rising sea levels can cause economic damage and population displacement, particularly in small island states—Bahrain above all—and low-lying urban areas of the Gulf, including man-made islands and land reclamation projects in Dubai and elsewhere. Estimates of the levels of rise vary greatly: according to the IPCC's report of 2007, the expected range is 0.18–0.59 metres by 2100, but as much as 10 metres or over if glacial melting is included.[34] Bahrain has repeatedly expressed its concern over the impact of rising sea levels. A national study published in 2005 predicted a 5–10% loss of territory (36–69 km²) by 2100 for a sea level rise of 0.2–1.0 metres.[35] A more recent study on the UAE, which has a considerably larger territory, predicts land losses of 1–6% (1,555–5,000 km²) by 2100.[36] In addition, according to the IPCC, extreme weather events like hurricanes and heat spells are expected to become more frequent as global temperatures rise.[37] In 2007 Gonu, the strongest tropical cyclone ever recorded in the Arabian Sea and the strongest ever to hit the Arabian Peninsula, struck Oman causing around 50 deaths, US$1bn worth of physical damage, and US$200m of losses in oil exports due to a production break.[38]

It has also been argued that climate change can precipitate existing social and economic problems in the Middle East, including inter- and intra-state tensions over scarce resources, migration induced by declining returns in agriculture and seawater intrusion (climate refugees) and, if the state's adaptation capacity fails, even increased poverty, unemployment, social instability and radicalization.[39] Poor states and weak governments are expected to suffer the most, as these have the lowest adaptation capacity. The GCC states, in turn, are relatively very wealthy; their state apparatuses are robust; they are not expected to suffer any major environmental crises in the near future; and they are not engaged in resource use-related disputes. Raouf, however, has noted that the lack of studies on the shared water aquifers in the Gulf is dangerous as, in a situation of scarcity, disputes may arise between the states.[40] A scenario in which water scarcity could cause internal tensions in the Gulf monarchies is, arguably, unlikely as long as the rentier bargain with the local population is upheld, as the governments have signalled low tolerance towards any demonstrations by non-nationals—and since 2011 by nationals.[41]

The issue of economic impact

It is difficult, if not impossible, to predict the economic consequences of measures to mitigate climate change, since there is high uncertainty even regarding the physical consequences. The widely cited but also criticized report by Lord Stern in 2007 estimated that greenhouse gas emission cuts, aimed at preventing a two-degree rise in temperature, would cost 1% of global GDP, while the consequences of inaction would cut it by 5–10%.[42] Although the Stern Review is at most only a best estimate, climate change mitigation and adaptation are deemed to require huge investments. The transition to a low-carbon economy will also create indirect profits and losses. Despite the lack of concrete evidence of such losses, as noted, potential economic losses induced by climate change *mitigation* have had the most weight in GCC OPEC member states' considerations of all the possible consequences.

There are four possible interlinked ways in which international climate change mitigation could negatively affect the GCC states' economies. Barnett and Dessai name three mechanisms that can affect oil export revenues: reduced demand, reduced price, and reduced market rent due to taxes.[43] If an ambitious global climate pact is agreed upon and implemented, these could all ensue. Furthermore, if global demand peaks before the six states' reserves are depleted, they will also suffer economic losses from unexploited resources.

Because of the importance attached by OPEC members to the issue of potential adverse effects of global climate change mitigation, or 'response measures', these have been studied in a number of model scenarios. The picture the studies present is mixed, to say the least. A scenario study conducted by Libya's former top oil official Shokri Ghanem *et al.* in 1999 predicted that if abatement targets under the first commitment period of the Kyoto Protocol (2008–12) were met, OPEC member states would be likely to suffer substantial export revenue losses. By the early 2010s, however, the scenario's basic assumptions had become badly outdated.[44] In a 2007 study, Persson *et al.* surveyed model-based literature from the late 1990s to the mid-2000s, which predicted that OPEC will lose up to a third of export revenues compared with baseline revenues by 2050 as a result of international climate politics (either carbon tax or cap-and-trade). The authors also developed a model that estimates potential OPEC revenue losses in a climate regime with universal emissions reduction targets, and discovered that this might actually increase OPEC revenues from conventional oil.[45] Results from peer-reviewed literature remain mixed, to say the least, and therefore conducive to polarized interpretations.

The OPEC countries themselves most frequently rely on the OPEC World Energy Model, developed in the 1980s and updated annually, to justify their case for the need for compensation. In the early 2000s, the model predicted the largest proportional losses in GDP by 2010 for the four GCC OPEC members (over 3% in Qatar and the UAE and 2% in Kuwait and Saudi Arabia), Iraq and Libya.[46] In addition to different calculation methods, different sides of the debate also present the figures in ways that support their positions: in 2009, the OECD's International Energy Agency, representing countries which are primarily energy importers and most of which advocate ambitious climate change mitigation, launched a study which projected that, under a 450 ppm scenario (which might limit the global temperature increase to below 2°C), OPEC's total oil revenue would be US$23 trillion in 2008–30. This is over four times higher than revenue during the previous two decades. Making comparison with a business-as-usual scenario, the IEA study stressed there would 'only' be a 16% total loss for the oil exporters. The Saudi negotiator Mohammed al-Sabban immediately dismissed the figures in the press as biased and referred to an older study by Charles River, which estimates annual losses of US$19bn from 2012 onwards for Saudi Arabia alone.[47]

As the models have so far demonstrated, predicting the demand and price of oil is extremely difficult, as these are influenced by a large number of other factors besides mitigation policies. The 2000s passed with no response measure-induced losses for the OPEC members; the 2008 price collapse was caused by the global financial crisis, not abatement policies. Moreover, in the early 2010s the swift creation of a global carbon market or implementation of carbon taxes seemed increasingly unlikely as the major emitting countries, the US and China in the lead, continued to display a lack of political will to commit internationally to ambitious emission cuts and targets. Furthermore, as Barnett *et al.* have reminded, payment of compensation for lost oil revenues, sought by OPEC, is 'politically unrealistic and practically problematic', because the calculation of the exact amount of losses is technically impossible.[48] OPEC countries will therefore need to look into other kinds of 'adaptation strategies'. What is more, the already existing domestic demographic pressures and natural resource security trends might well push the Gulf monarchies to economic diversification away from oil far earlier than global mitigation would.

The double challenge of adaptation

According to some studies, physical impacts of climate change are already visible in the Lower Gulf. Most impacts, however, are expected to take place in the coming decades. By then, fossil fuel-derived external rent is expected to be dwindling as a result of resource depletion (particularly in Bahrain and Oman) and/or declining global demand (new technologies or international climate change mitigation).[49] In addition to this, regional instabilities can have destabilizing impacts in the otherwise internally robust Gulf monarchies: climate change-induced drought, poverty, unemployment and migration can turn the now turbulent neighbourhood into one of failing states and large-scale social unrest. As in the case of environmental change in general, those countries best equipped in terms of functioning institutions, financial resources and good governance will be in the best position to cope with the threats and challenges associated with natural resources and climate change. Adaptation has not been a major concern for the Gulf monarchies. Awareness of projected climate change impact and its links to the states' natural unsustainabilities has only recently begun rising. Measures that environment researcher Raouf called for in 2008—increased data monitoring, public awareness raising, mainstreaming of climate change impacts into policy-making, and policy implementation—had only begun appearing in government rhetoric in some of the GCC states three years later.[50] Abu Dhabi emerged as the pioneer with a number of studies on the impact of climate change, a major awareness campaign on natural resource use and climate change, and important policy targets and instruments in place as of 2012. Intra-GCC cooperation in the area of adaptation is also lacking. However, many existing policies and reform processes are likely to enhance the monarchies' adaptation capacity: the most important being economic diversification; resource conservation campaigns and the impending subsidy reforms; and sustainable energy and water management practices and food security policies.

There are also potential adverse effects of national mitigation measures that affect any government's willingness to act. Buhaug *et al.*, for example, have argued that strict measures to restrict CO_2 emissions in high-growth developing countries would most probably damage their economic growth, which could result in political instability and civil unrest.[51] For a Gulf rentier government this is a serious issue, particularly when it comes to natural resource subsidies. On the other hand, as the Stern Review reminds us, delayed action might cost more than prompt action.

Indicators of natural unsustainability

Two indicators, carbon dioxide emissions (per capita) and the aggregate ecological footprint index, serve for measuring the Gulf monarchies' unsustainability with regard to their natural resource consumption and the environment. The former is a good indicator of energy use-related unsustainability within a country's borders: it includes the impact of the energy industries, which, importantly, adds up to the per capita emission levels in the GCC states. Data on carbon dioxide emissions does not give a complete picture of the problem: Qatar's CO_2 emissions in 2005 were estimated at 45 Mt, but total greenhouse gas emissions amounted to 61 Mt of CO_2 equivalent (CO_2e).[52] However, CO_2 is the most important single contributor to the problem and general availability of data for this is much better and broader.

Both historically and currently, the GCC states' total CO_2 emissions are small. According to the World Resources Institute, the cumulative historical CO_2 emissions of the six states from 1850 to 2007 represent 0.04%–0.58% of the world total. The states' total emissions in 2007 (2.4% of global total) amounted to only a fraction of US or Chinese emissions (19.7% and 22.7%) in 2007. Their per capita emissions, however, are the highest in the world, as is shown in table 8.

Table 8: Carbon dioxide emissions of the GCC states in a global context, 2007.[53]

	CO_2 emissions (Mt)	CO_2 emissions, % of world total	CO_2 emissions, global rank	Per capita emissions of CO_2 (tonnes)	Per capita CO_2 emissions, global rank
Bahrain	21.4	0.07	80	28.1	3
Kuwait	69.7	0.24	46	26.2	4
Oman	40.3	0.14	67	14.8	12
Qatar	55.6	0.19	56	48.8	1
Saudi Arabia	373.4	1.26	18	15.5	11
UAE	138.4	0.47	33	31.7	2
GCC	698.7	2.37	–	19.5	–
MENA	2,216.9	7.51	–	4.8	–
Non-Annex I	14.489.9	49.07	–	2.8	–
United States	5,826.7	19.73	2	19.3	7
China	6,702.6	22.70	1	5.1	66
World	29,259.1	100.0	–	4.5	–

Owing to the growth factors described in the sections above, the GCC states' emissions are estimated to have grown rapidly in the past decades: from 1980 to 2007, their CO_2 emissions from energy use grew by an average of 5.5% annually, compared with −0.3% in the EU-27 and 0.8% in the US. The emissions of the MENA region, which accounted for 7.5% of the global total in 2007, are projected to grow faster than the global average in the coming decades. Despite the uncertainties associated with emission projections, as with energy demand projections, they provide a useful general indication: the French POLES model estimates the Gulf states' average annual emission growth as 2.7% in 2007–2030, while placing global average growth at 2.2% and that of the rest of the Middle East at 3.4%. Emissions in most European countries and the US are projected to grow by less than 1% per annum.[54]

The high per capita emissions of the Gulf monarchies, which placed them as the top emitters in the world in 2007 (ranks 1–4 and 11–12), has attracted unwanted international attention since the late 2000s. The Western press has published articles on the unsustainability of the Gulf societies,[55] and pressure has been mounting in the UN climate change negotiations for the GCC states (alongside other developing countries) to take on new commitments. In 2007, according to the World Resources Institute, per capita emissions in Qatar were almost 11 times the global average and even 2.5 times the US average (see table 8). As noted, however, after taking into account the emissions of the industrial sector and the marked differences in energy consumption patterns within the populations, it becomes apparent that per capita emission is actually 'no-one's emission'.

A better indicator for measuring consumer lifestyles and the impact of the non-oil part of the economy broadly is the ecological footprint index, which only counts the resources consumed and waste and pollution produced within the borders of a country, including imported products like food and excluding products like oil and gas that are exported. The footprint, which is measured per capita and expressed in 'global hectares' (gha), covers all countries with a population of over 1 million (which so far has excluded Bahrain). The latest report, which includes data for 2008, ranks Qatar (11.7 gha), Kuwait (9.7 gha) and the UAE (8.4 gha) as the countries with the largest footprint in the world, ahead of high income OECD countries like the United Kingdom, United States, and Japan. The authors of the footprint index, including WWF, calculated that the world's overall biocapacity (or what nature can regenerate each year) in 2008 was enough to support a footprint of 1.8 gha per capita only. This means that the three GCC states' residents *on average* lived like there were five to seven planet Earths. Given the high share of extremely low-footprint migrant labourers in these two fast-growing and building economies (57% of Qatar's population in 2008), it is obvious

that the ecological footprint of the wealthiest segment of the populations, consisting of upper middle class expatriates and local citizens, is
still several times higher. Additional problems are the cruel heat, lack of
sufficient renewable water resources, and arid climate, which translate to
lower than average biocapacity in three of the five states included in the
analysis (0.4–0.7 gha/capita). Only Qatar and Oman rise slightly above
the average (2.1–2.2 gha/capita) owing to higher scores in fishing grounds,
but even these two are still left with an 'ecological deficit' of 9.6 and 3.5
gha/capita respectively.[56]

Contemporary societies in the Gulf inflict a heavy toll on the surrounding environment, not only through their greenhouse gas emissions. An
extensive study from 2006 by the Gulf Research Center on the state of
the environment in the GCC states listed six pressing environmental
issues. Besides increasing water scarcity, the most pressing challenge,
according to the report, is land degradation and desertification, caused by
population growth and urbanization, overgrazing, and intensification and
expansion of agriculture. Marine biodiversity is constantly endangered by
oil spillage, human settlements in coastal areas and seawater desalination.
Air pollution is also considerable because of high per capita carbon dioxide emissions. Finally, solid waste management is becoming increasingly
problematic with population growth and high consumption patterns and
very limited recycling.[57] The emirate of Sharjah, for example, produces
3.4 kg of waste per capita each day.[58] Indeed, the energy, water and food-
related unsustainabilities discussed in this volume are not the only negative externalities of uncontrolled growth in the contemporary GCC
rentier states, but arguably they will pose the greatest challenges for the
Gulf monarchies' business-as-usual development trajectories, as well as
the local governments in their quest for regime survival.

Mitigation

Policies and actions officially labelled as national mitigation measures are
still very few in the Gulf monarchies, as the six states are classified in the
climate convention as developing countries and hence do not have any
obligations under the Kyoto Protocol to cut or limit emissions before
2020. A Chatham House report describes mitigation activities in the
OPEC member states up to 2005 as the result of wider developments in
the energy field. These include investments in more efficient technolo-

gies; the development of gas markets, investment in non-associated gas and recovery of associated gas; and technological developments, including gas flaring recovery and liquefied natural gas (LNG).[59] Notably, however, even in the Western countries the most substantial emission cuts have always been motivated by energy security concerns rather than a preoccupation with climate change.[60]

While domestic energy security and improvements in fossil fuel technologies will most likely be the main drivers of mitigation in the GCC, technology transfer and market mechanisms, like the Kyoto Clean Development Mechanism (CDM), will play a major role in supporting this trend and also encouraging projects in other areas that lack prospects of major short-term economic viability or are otherwise 'unsexy'. These include projects in the areas of energy efficiency, renewables, waste-related emission cuts and energy production. The CDM is one of the flexibility mechanisms under the Kyoto Protocol. Its purpose has been to advance sustainable development in developing countries and assist developed countries to meet their emission targets. With only two notable exceptions, the Gulf monarchies have been slow to seize the opportunities of the CDM. These have been Qatar's massive gas flaring reduction project from 2007 and a number of energy projects (solar, landfill and efficiency) in the UAE starting from 2009, including a 100 MW solar thermal power plant in Abu Dhabi. In 2011, a large gas flaring reduction project in Oman and two landfill gas capture projects in Saudi Arabia also sought registration under the UNFCCC.[61] Prominent developments in the area of alternative energies and technologies since 2006 include Abu Dhabi's Masdar Initiative and projects undertaken under the broad umbrella of the Qatar Foundation, in particular linked to its Qatar Science and Technology Park.

By the end of 2011, the most concrete mitigation target announced by a Gulf monarchy had been Abu Dhabi's declaration of 2009 to reach a production capacity share of 7% for renewables by 2020. Other monarchies too have since floated targets, but these have not yet been coupled with implementation plans. Also, clear carbon (intensity) reduction targets are still missing, and emission registries are scarce and largely of bad quality. Policy measures, including tax incentives and feed-in-tariffs for low-carbon energy technologies, are also largely lacking, and climate change is only slowly beginning to be taken into account in planning in the different policy sectors.[62] By 2011, carbon reduction targets had not yet been announced by any of the six states, and the political realities made it seem unlikely that any targets would be launched in the coming years; the international climate process was stagnating and ambitions were low so there was little external pressure on the GCC states to make new international commitments. Emission data, on the other hand, have been improved significantly in the UAE, and Qatar too has published a national registry for 2007. Still, bureaucratic competition and opacity continue to hinder data availability and reliability. Feed-in-tariffs for solar energy have been discussed in the Gulf since the late 2000s, and climate change considerations began appearing on water and food security policy agendas. Few concrete actions, however, had been taken as of the early 2010s.

The external impact: rising image awareness

The different forms in which Gulf monarchies' governments adopted climate change on their policy agendas from the late 2000s largely correlated with a general rise in their awareness about the increasing environmental unsustainability of prevalent domestic development patterns. External image perceptions, however, often played a role in these developments. A case in point is the degrading impact of the fast and largely uncontrolled construction activities of the 2000s on the local environment, which became a source of international attention and, partly because of this, has finally begun receiving due attention.

The first one to respond was Dubai, at the height of its economic boom. Luxury lifestyles had become one of the cornerstones of the emirate's branding strategy, and the government had grown increasingly sensitive to bad international press that could harm this modern image. The UAE had figured at the top of the ecological footprint index since its onset in 2000, but according to close observers, it was the 2006 index that prompted Dubai's ruler Sheikh Mohammed to react. That year, the UAE's ranking was covered in Western media, with mentions of the high ecological impacts of Dubai's indoor ski slope and the lavish water use patterns. In initial denial, a government spokesperson declared that data on population and re-exported goods were flawed.[63] Soon afterwards, however, the government teamed up with the local branch of the global environmental organization WWF to recalculate the country's footprint. The Al Basama Al Beeiyah (Ecological Footprint) initiative resulted in many unexpected consequences, including an unprecedented nationwide data collection effort, and, prompted by the results of the study, a major awareness campaign aimed at curbing energy and water usage in households, as well as a new study on potential future emission scenarios. While the new calculations did not change the UAE's international ranking, they were received in a constructive manner, and led the governments in Dubai (and Abu Dhabi too) to recognize the positive synergies between curbing natural resource wastage, energy security, and improving the federation's external image.

In 2007, Sheikh Mohammed of Dubai also announced a green building code to be enforced in all buildings constructed in the emirate starting from 2008.[64] The economic crisis, among other factors, meant that enforcement would be repeatedly postponed to the near future, on economic grounds: 'the Government does not want any additional costs to building owners and investors',[65] said a utility sector executive in 2011. While much of the reluctance towards sustainable building in the industry stems from short-term price considerations and the lack of required specific skills among the existing labour force, a lot of the inertia in the green building sector stems precisely from the lack of regulations and policy signals—and the omnipresent high fossil fuel subsidies.

For Abu Dhabi too, low scores in international environmental comparisons triggered action. This happened when the emirate recognized that lack of relevant national data was placing the UAE in the second worst category and listing it as the worst GCC state in environmental policy goal performance (at rank 152 of 163) on the American Environmental Performance Index.[66] In 2008, the local environmental agency EAD sought methodological advice from the authors of the index at Yale and Columbia Universities. The result was the development of an emirate-specific environmental performance index to collect data from areas specifically designated as important for the local context. The UAE's Foreign Ministry also seemed to consider the country's bad environmental record as an image issue: in an internal review in 2010, the Ministry included renewable energy and climate change in the areas in which the government should 'explain [its] position and represent [this] position better'.[67]

Not all GCC states, however, shared this heightened awareness of and concern over their international green record. As will be discussed in later chapters, even in 2011 intransigent climate policy positions, and often little more than lip-service to nation-wide improvements in the area of environmental sustainability, revealed how the other three wealthy rentier states, Qatar, Kuwait and Saudi Arabia, did not yet consider environmental image a top foreign policy priority.

Impact of the political system: authoritarianism and public awareness

Societal forces, such as interest groups, institutions, public opinion and media, had a mixed record in driving energy and environmental policies in the Gulf in the 2000s. Underpinning all social interactions is the fact that the political systems of the Gulf monarchies are classifiable as authoritarian, or semi-authoritarian at most. As already shown in table 3, in Freedom House's Freedom in the World index for 2010, all but Kuwait (classified as 'partly free') were ranked as 'not free'. The Economist Intelligence Unit's democracy index categorized all six states as authoritarian, with Kuwait again the least authoritarian and Saudi Arabia and the UAE the most.[68]

The overall ability of the GCC states to avert political reforms in the past decades has been relatively strong owing to the combination of abundant rent allocations and the skilful application of immaterial legitimacy resources and strategies towards the national populations. Typically, the impetus for reforms has been the transition of power to a new member of the ruling family. These gradual reforms, as argued by Nonneman, have been used for 'political decompression', aimed at creating 'liberalised autocracies' at most.[69] The GCC governments' reactions at home to the revolutions in Egypt and Tunisia in early 2011 were almost exclusively counterrevolutionary and centred on heavy allocations of wealth and welfare benefits, as described above.

Reflecting a broader pattern of silenced, co-opted or inexistent civil society organizations, environmental non-governmental organizations in the GCC are few. They mostly rally around non-controversial, low-impact-high-visibility themes like beach clean-ups, tree-planting, recycling and periodical awareness-raising events and lectures. This proves that compulsory licensing and co-opting through state funding have proven to be successful mechanisms, from the governments' perspective, for controlling and monitoring bottom-up interest aggregation in environmental issues.

Authoritarianism has a number of consequences for environmental governance and policy-making, since the concentration of decision-making powers in the hands of a small political elite can function either as a hindrance or as a catalyst for action. If an issue is perceived to be in the interests of a top decision-maker, the launch of a policy or plan and its implementation can be very fast and effective. In the Gulf, concrete examples of this are Qatar's LNG programme since the late 1990s, the Dubai Model (as articulated for example by Hvidt) pre-2008 and Abu Dhabi's nuclear energy programme, launched in 2008. In an opposite case, a lack or loss of interest on the part of top leaders can lead to grandiose visions remaining as plans, as has happened with Dubai's green building legislation, or suffering important delays or downscaling, as in the case of Abu Dhabi's Masdar City from 2009 onwards. Alternatively, announced plans and policies with withering political support have disappeared within government institutions, such as Qatar's renewable energy strategy, planned at least since 2009 and delegated in 2010–11 to a small unit within Qatar Petroleum; and a new domestic energy policy by Abu Dhabi's government, first announced in 2009 and still unpublished as of 2012.

The role of institutions in the GCC monarchies in shaping outcomes should, however, not be underestimated. For example, in Saudi Arabia and Qatar, powerful Energy Ministries with symbiotic links with national oil companies have, until recently, played an important role in both shaping international climate policy positions and sidelining renewables on the domestic energy agenda. Furthermore, as Gulf nationals often note, the best way to kill a stated goal is to establish a committee around it. Indeed, this fate seems to have befallen at least the UAE's climate policy, under preparation since at least 2009, and delegated to the UAE's new National Climate Change Committee in 2010 (discussed in chapter 4).

Equally important in bringing about changes in government attitudes is the potential of public opinion and, in the case of the GCC states, that of citizens. Even in monarchies where the top decision-making elite enjoys little resistance and high autonomy, such as Qatar, the ruling bargain is such that the leadership still has to give heed to the values and moral codes of the local majority, in particular when dealing with religious issues. In the case of environmental values, available polls indicate rising awareness among both residents and nationals. A three-year survey in Abu Dhabi registered considerable rise in awareness (20 percentage points) and sense of responsibility for protecting the local environment (40 percentage points) from 2008 to 2010.[70] In a 2010 World Value Survey conducted among Qatari nationals, 98% of the respondents identified themselves with the statement 'it is important for this person to be concerned for the environment and for protecting it'.[71] When it comes to observable behaviour, however, any Gulf observer can easily identify a large amount of factual evidence indicating the opposite, ranging from cars of the size of small lorries and palace-size villas, litter, waste of food and water and heavy consumerism, to low participation in environmental organizations and campaigns. Simultaneously, a large bulk of the Gulf resident populations is confined to living more ecological lifestyles because of low income and not as a lifestyle choice. There is no doubt that a considerable proportion of those nationals and expatriates who can afford to choose, place material improvements in personal living standards, such as bigger cars and houses or heavy use of air conditioning, ahead of 'higher goals' of environmental protection. Simply, if this were not the case, GCC states would not have the world's highest ecological footprints.

The issue of environmental awakening in autocracies is particularly problematic as environmental policy-making and actual implementation and follow-up of more sustainable practices have largely become the victim of a chicken and egg situation: on the one hand, without an environmentally aware population the government lacks incentives to take action that could shake the status quo; on the other, lack of official channels for civil society participation effectively discourages the formation and continuation of interest groups that could pressure the government to action. The ubiquitous unwritten prohibition against open criticism of top elite members, institutions and key policy elements, common to most GCC monarchies, further discourages environmental groups from building agendas that address 'serious issues' like the environmental damage caused by industries and construction activities, or making critical comments on sensitive areas such as nuclear energy. In the case of the UAE, for example, residents openly critical of the country's emerging nuclear programme are known to have been told off, even dismissed.[72] It is no coincidence that Greenpeace, for example, does not have an office in any of the GCC states; in contrast, as close by as Jordan, a local activist group has since May 2011 led an active anti-nuclear campaign, and further west, Lebanon boasts a number of active environmental civil society groups with well-defined agendas that strike into the core of climate policy-making and have active working relationships with relevant government institutions.

The transformative power of the media is also heavily restricted in most GCC states, as the press, consisting in many cases of expatriate journalists, faces both written and unwritten (in Qatar, for example, there is no media law) and true and imagined restrictions on content and style and, as a consequence, exerts heavy self-censorship. Although environmental issues mostly fall in the category of harmless, even the occasional products of investigative, critical environmental journalism steer clear of big issues like air pollution from the local energy industry or the environmental impact of land manipulation. As was openly admitted in early 2011 in an unusual editorial in *The Peninsula*, a Qatari newspaper, harassment of journalists is common and often arbitrary, and 'and even if a journalist shows guts and is willing to do critical or informative writing, he has no access to officials to get basic information'.[73] Access to basic information on government decisions, policies and procedures, even laws, is a further issue that affects not only journalism but also researchers into the region's states. From a government and local big business perspective this state of affairs is convenient, as it is difficult to build a strong case against something of which there is no tangible evidence or data.

Although environmental awareness among the local, largely Western-educated youth is arguably rising, this has not yet translated into visible collective action among the citizen population.[74] Some exceptions exist, such as Green Jeddah, a group of largely female students who seek to raise awareness about the Saudi city's local environmental degradation. Although often hard to pinpoint, it can be argued that a lot of the 'greening' that has been taking place in governments' discourse in the Gulf in the past few years is targeted at the youth who form a high proportion of these states' citizen populations. According to some estimates, in Saudi Arabia 60% of the total population are under 21.[75] New, green role models for Gulf citizens have also begun emerging alongside a few pioneers. The 'older generation' includes figures like Habiba al-Marashi of the Emirates Environment Group in Dubai and Saif al-Hajari of the Friends of the Environment Centre in Qatar. Newer role models include Abdulaziz Al Nuaimi, the 'Green Sheikh' of Ajman; Omran al-Kuwari, a Qatari solar entrepreneur; Abdulla al-Missnad, a Qatari who travelled to Antarctica (together with the Green Sheikh and other Emiratis) to raise awareness of climate change; Haidar Taleb, an Emirati who travelled through the UAE in a solar-powered wheelchair; and the 10-year-old Abu Dhabi resident Abdul Mugeeth, known as the Paper Bag Boy, who raises awareness by reusing newspapers.

Non-governmental organizations: influence from within or without?

Although this book largely focuses on government-related, top-down responses to climate change and natural unsustainability, consciousness of what is happening in the society is crucial, as non-profit groups and private sector companies, particularly those with international origins or headquarters, have in some areas been in the forefront of the 'green agenda' in the GCC. In the cases of Abu Dhabi and Qatar, semi-governmental entities with important international business and staff links have not only participated in but effectively led the push in areas like R&D and education in the area of alternative energy technologies and environmental sustainability; implementation of renewable energy; sustainable building codes and projects; and promotion of natural sustainability-enhancing policies on the government agenda.

In the two monarchies examined in this book the fossil fuel sector, somewhat counter-intuitively, has been among the pioneers of environmental protection and projects. Owing to the numerous joint ventures in this sector, however, it is only natural that international regulations should become incorporated in the corporate practices of national oil and gas companies like Abu Dhabi's national oil conglomerate ADNOC, Qatar Petroleum, RasGas and Qatargas. While the companies commonly give assurances that their environmental impact assessments, regulation, monitoring and enforcement are of international standard, the companies' opacity makes it difficult for independent observers to verify these assertions. As described by an environmental official in the UAE, ADNOC resembles and behaves like a 'kingdom within a kingdom'.[76] The same can be said of the other case study of this book, Qatar, and its own kingdom of Qatar Petroleum. Over the past decade, environmental sustainability has received more attention in private sector corporate social responsibility projects in the region. Both oil sector-led and private company-led environmental projects face little scrutiny, however, owing to the toothless character of the local environmental 'civil society' and media. Greenwash is prolific and big polluters get away with small, often unrelated offsetting projects or one-off events, such as the corporate membership of the Emirates Wildlife Society or Clean Up Arabia, an annual beach clean-up day organized by the Emirates Diving Association and supported by various national oil and gas companies of the region. Companies' offsetting initiatives and support for local environmental groups are naturally a double-edged sword, as they bring in important funds for often small local groups. As pointed out by a Kuwaiti environmental consultant in 2011: 'all these companies are not going anywhere, so we may as well try and do something positive together'.[77]

The GCC monarchies' high international exposure to globalization, and especially those with high shares of foreign populations, means high exposure to prevalent global trends in areas like policy, business and values. The rise of climate change on the international energy agenda has prompted a wave of coat-tailing trends in industry and finance, which GCC companies and governments have been quick to take advantage of. Among the most prominent of these trends are the second rise of the nuclear industry, and the green building, renewable energy and clean technology sectors. Governments have led the way in investment, starting with the four major GCC states' pledge in 2007 of US$750m to an OPEC clean tech fund. More generally, Western-born ideas of ecological or green lifestyles are brought into the GCC states in a number ways, including US- and UK-educated young GCC citizens, international television channels, and Western (-minded) expatriates who often occupy key advisory positions in government-related institutions and large companies, or consulting companies offering services to these. While the impact of these people and mechanisms on the 'greening' of Gulf monarchies' policy agendas is hard to measure, the central case studies in this book—of Abu Dhabi's Masdar Initiative and its civilian nuclear energy programme; the Qatar Foundation's sustainability projects; and Abu Dhabi's bid to host the International Renewable Energy Agency—provide several examples of the ways in which international trends flow in and out of these dynamic and ambitious post-traditional desert autocracies.

The illusion of abundance

Oil is a relatively short-term phenomenon in the harsh environment of the Arabian Peninsula: the first to begin production was Bahrain in 1933, and Oman was the last in 1967. Although it has dominated and revolutionized the monarchies' economies and politics throughout their independence, production has already peaked in Bahrain (1970s), Dubai (1991) and Oman (2001).[78] Despite the pronounced differences in reserve futures, the fast and deep penetration of oil (and increasingly natural gas) into the modern societies it has helped to create has produced dependence so thorough that governments have seen it as most convenient to maintain an illusion of abundance, no matter how short-sighted this might be.

As has been described in this chapter, it was this illusion that in the 2000s began to be challenged by both internal and external resource use-related pressures, dynamics and developments. In most GCC monarchies, economic and population growth in the 2000s was so fast that governments, infrastructure, policies and the local environment struggled to keep up. Even though the 2008 financial crisis brought a badly-needed pause moment, from the perspective of re-evaluation of many exaggerations of the preceding boom years, the pause for natural resource demand growth was barely visible: after doubling their existing power generation capacity between 2002 and 2009 (by 50 GW), according to a World Energy Council estimate the GCC states' capacity will again need to double (by 100 GW) over the present decade, at a cost of over US$300bn.[79] Simultaneously, future demand estimates for oil are being complicated by the uncertainties of the global transition to low-carbon. Climate change also brings a set of physical challenges, which are expected to interplay with the monarchies' existing natural unsustainabilities.

Fossil fuels, related revenues, subsidies and emissions, together with the political economies and ruling bargains that evolve around them, are at the core of the growing natural unsustainability that has come to menace the business-as-usual ways of going about development of the contemporary Gulf monarchies. Since the late 2000s, as climate change and the need for increased natural sustainability have crept into the local governments' consciousness, many promises of improvements in this area have been made. Almost equally many have not yet been duly implemented. Despite the many similarities among the Gulf monarchies, there have been marked differences in responsiveness, primarily related to the strength of the rentier state and elite autonomy, but also driven by a number of additional structural and human agency-specific factors. The following chapters will examine the natural sustainability and climate change-related responses of two of the most active governments, those of Abu Dhabi and Qatar, and analyse the drivers behind these responses.

4

ABU DHABI'S NATURAL
SUSTAINABILITY COMPLEX

Since 1971 the emirate of Abu Dhabi, the sovereign owner of more than 90% of the UAE's fossil fuel wealth, has sustained the federation's unity and economic development, backed by a combination of external rent and skilful leadership. Behind the success story, however, is the dark side: the oil-based economy, a resource-squandering rentier bargain, high per capita consumption figures, and a diversification strategy reliant on energy-intensive development patterns. Together, these factors have contributed to a development trajectory that is increasingly unsustainable, environmentally and natural resource-wise. While oil remains abundant and the rentier state strong, during the past decade Abu Dhabi has undergone profound transformations in its leadership, development planning, and governance structures. The emirate has felt the same domestic pressures that confront the other GCC states, including high population and resource demand growth, but there is a further layer of challenges it faces as the leader of the loose seven-member federation. Furthermore, Abu Dhabi, as a major oil producer and member of OPEC, has a key stake in the current global energy paradigm. As a key difference from its OPEC peers, however, the emirate's leadership has been capable of seeing beyond the oil era and has devised some of the most innovative domestic approaches and proactive responses to environmental unsustainability and climate change in the region to date.

For understanding the multiple drivers of the recent natural sustainability developments in Abu Dhabi, the book takes a dualistic conceptual approach. First, this chapter lays out the broader, underlying context, through an analysis of what can be described as a 'natural sustainability complex'. The examination covers the political economy of natural resources and the structures and dynamics of environmental and climate change-related decision-making in the emirate of Abu Dhabi, with regard to its relations with the other emirates. As a strong example of the power of human agency, the chapter also discusses the role of Sheikh Zayed and his sons in shaping Abu Dhabi's institutional context and driving present policy choices.

After this, chapter 5 examines more closely the role of human agency and decision-making dynamics in shaping Abu Dhabi's development and policy outcomes relating to natural unsustainability and climate change in the late 2000s. It starts with an overview of the physical and economic challenges of climate change and resource scarcity to Abu Dhabi and the UAE, briefly describing the broader government responses. The main part of the chapter consists of two detailed analyses of the drivers and dynamics of alternative energy decision-making, presented by case studies of the Masdar Initiative and Abu Dhabi's nuclear programme. Going beyond the domestic structure-agency axis, the chapter also shows how external factors, such as the intensifying impacts of globalization and historical relationships of dependency, contribute to decision-making outcomes in Abu Dhabi and the broader federation.

Chapters 4 and 5 together demonstrate the undeniable significance of domestic structures and key actors. But they also show how external factors, such as existing foreign policy priorities and alignments, the role of external allies and trade partners, and the politically motivated quest for prestige and economically motivated branding strategies, can play a role in domestic-level government responses to climate change and natural unsustainability. They also highlight the opportunities and challenges arising from a system that still depends to a large extent on neotraditional forms and channels of rule and decision-making, and is characterized by a mixture of young and old, weak and powerful decision-making institutions, which face numerous challenges relating to human capacity, competition and coordination, among others.

Fossil and non-fossil economies

As established in the previous chapters, Abu Dhabi is a strong rentier state with a strong material ruling bargain. Despite an ambitious economic diversification agenda, its government seeks to extend the status quo to the far future. Since the 1980s, the main development objectives and challenges of the federation have essentially remained the same. Led by its two complementing economic motors, Abu Dhabi and Dubai, the UAE government has aimed to sustain high growth, diversify the economy away from oil to stable non-oil income sources, and gear the national population towards productive labour.[1]

Economic growth, on average, has been the most successful aspect of the UAE's development: in 2010 the economy ranked 36th in the world, and the country's GDP per capita was among the top 10.[2] Marked socio-economic differences, however, exist between and within the seven emirates. Abu Dhabi's share of the federation's GDP is around 55–60%. Its GDP per capita (US$73,000 in 2007) is several times larger than that of the poorest emirate, Ajman (US$12,000 in 2007), which sets Abu Dhabi's nationals, on par with Qataris, among the richest people in the world.[3] Abu Dhabi is the sovereign owner of 94% and 93% of the federation's proven oil and natural gas reserves (92.2 billion barrels [bn bbl] and 5.6 trillion cubic metres [tcm] respectively).[4] Dubai's limited oil production peaked in 1991 and the reserves of the five smaller emirates are even less significant.[5]

According to government data, the UAE's economy as a whole was the least rent dependent of all GCC states, with crude oil and natural gas accounting for 31% of GDP in 2010.[6] The importance of fossil fuels in Abu Dhabi's economy, however, is still markedly high. Oil was reported to have constituted 45–59% of GDP in 2005–10,[7] but the proportion could be even higher as part of oil revenues are paid directly to reserve accounts.[8] According to the International Monetary Fund, over 80% of Abu Dhabi's income derives from the national oil company ADNOC and its fourteen group companies. In July 2008, at the peak of oil prices (US$147/bbl), with oil production at 2.5 million bbl, the emirate was estimated to earn US$351m a day from oil and gas.[9] The implications of the high dependence become even more tangible when viewed against total hydrocarbon revenues: from 2003 to 2008, the value of Abu Dhabi's oil and natural gas exports tripled from US$23bn to US$94bn, after which it dropped to US$60bn in 2009, rising again to US$68bn in 2010 with the recovery of global oil prices.[10]

Owing to its wealth and comparative advantage brought by its oil reserves, Abu Dhabi's long-term economic strategy relies extensively on a combination of overseas investments and industrialization in energy-intensive industries (such as aluminium and petrochemicals) and the hydrocarbon sector. Industrialization began shortly after the discovery of oil in 1959, although export revenues were not used for infrastructure development until 1966, when Sheikh Zayed bin Sultan Al Nahyan assumed leadership. Up to his death in 2004, the emirate's development dragged behind Dubai's.[11] Even today, despite the dizzying pace of lavish diversification projects, the shift away from oil is planned to be gradual, since Abu Dhabi's oil reserves are officially estimated to last close to a century at current production rates.

The importance of overseas investment as a supporting survival strategy has grown significantly since the establishment of the Abu Dhabi Investment Authority (ADIA) in 1976, which, even after losses following the 2008 financial crisis, is still considered as the world's largest sovereign wealth fund, with US$627bn of assets in 2011 according to the SWF Institute.[12] In addition to ADIA, there are at least seven smaller funds either owned or controlled by the local government, including the US$13bn Mubadala Development Company.[13]

In 2004, with a new generation of leaders in charge, the government initiated a fast-paced diversification drive to combat inflation and skewed trade balances, and address the need to create jobs for the growing national population.[14] Accompanied by massive infrastructure developments and modernization programmes, the diversification drive soon materialized in a carefully thought-out selection of new economic sectors, including high tech heavy industries, cultural tourism, real estate and alternative energy technologies.[15] Characteristically, projects in these emerging sectors are designed and built with top global partners, so as to ensure the highest possible chances of success, leaving their dependence on top elite members' patronage as the only major contingency factor, as will be demonstrated.

Since 2004, Abu Dhabi began to emulate Dubai's model in the development of its real estate sector and diversification of its sources of (potentially volatile) external rent. Although the real estate sector also serves as an extension of the allocative state and in maintaining patronage links between the government and citizens,[16] Abu Dhabi's government still owns most land, and megaprojects continue to remain in the domain of powerful businessmen within or linked to the royal family.[17] In 2007, the total value of all planned and ongoing construction projects in the emirate was estimated at US$300bn.[18] Also, since 2005, land ownership rights have gradually been extended to non-nationals through long-term leases in order to attract foreign capital.[19] Abu Dhabi's free zones will also support diversification. The current and planned ones include the Industrial City of Abu Dhabi, Khalifa Port and Industrial Zone, and Masdar City, the former two of which accommodate industries of different sizes while the latter specializes in alternative energy technologies.

Expansion plans in the tourism sector are massive, with expectations of 4.9 million tourists by 2020 and 7.9 million by 2030, compared with 1.8 million in 2007.[20] The developments, many of which are linked to the expansion of the real estate sector, are aimed at luxury tourism, some incorporating nature conservation or local cultural heritage, others F1 racing or golf. The expansion of the tourism industry, expected to employ mostly foreigners, will be boosted by the 'national' airline Etihad's US$43bn fleet expansion.[21]

Projects in the area of alternative energy technologies have largely been concentrated under the umbrella of the Masdar Initiative. Founded in 2006 by Mubadala, Masdar has quickly established itself as the UAE's focal point of developments relating to R&D, foreign asset acquisitions, technology transfer, and development and implementation of new energy technologies, in particular renewables. If successful, the seeds planted by Masdar will contribute to building the local knowledge economy and provide jobs for nationals. In the future, with the rise of domestic renewable energy production and possibly carbon capture and storage, this sector also has potential to contribute to Abu Dhabi's rentier economy by saving fossil fuels for export or even producing external rent to government coffers, for example through solar energy exports.

Population pressures

Estimates of the UAE's total population have ranged between 5.1 million and 6.0 million in 2009 and 7.9 million in 2011.[22] As in some other GCC states, the country's demographic data are presumed to be a victim of politicization: owing to the great divergence between various agencies' figures, and suspected attempts to bloat numbers for prestige and image reasons, available data have become merely indicative. The federal structures further complicate coherent data aggregation. During past decades Abu Dhabi's population has grown rapidly, from an estimated

200,000 in 1975 to 950,000 in 1995 and 2.0 million in 2010. As a sign of unexpectedly fast growth, even despite the economic downturn in the late 2000s, the total population in 2010 had already reached a level it had been expected to reach by 2020 in Abu Dhabi's first long-term urban planning document, launched in 2007. Official estimates place the nationals' share of the total population at 22%. Similarly to the other Gulf monarchies, the emirate has a high male to female ratio (2.3 in 2010), resulting from the economic reliance on foreign labour in male-dominated sectors of the economy.[23]

Abu Dhabi's national population is young, and this presents a job-creation challenge to the local government: in 2009, 40% of the emirate's nationals were below 15 years old, and in 2008, 10.4% were registered as unemployed. Average population growth rates for nationals were 9.1% in 1975–85 and 4.5% in 1995–2005, whereas for non-nationals they were 10.1% and 3.8%, respectively.[24] The fast growth of the non-national population has brought a series of challenges, ranging from energy insecurity to dilution of the Emirati identity, and some nationals have expressed concern about the fast decline in birth rates among the native population.[25] In an attempt to increase the country's permanent knowledge base, the state seeks to guarantee all Emirati high school graduates a place in a university. Almost two thirds of university students are female, a ratio which is presumably reversed in the labour market.[26] Often demand in the local labour market and supply of qualified and willing Emiratis do not meet. In 2008, expatriates accounted for 99% of the private sector and 91% of the public sector jobs in the UAE. 63% of Abu Dhabi citizens worked in the public sector. By 2020, according to some estimates, Emiratis will constitute only 4% of the total workforce.[27] The government implicitly admits that Emiratization policies so far have largely failed.[28] The lower wages of the private sector, often low working morale, and negative stereotypes associated with nationals have been mentioned among the reasons.[29] In order to increase efficiency and cut direct costs, in 2007, Abu Dhabi's government embarked on a rationalization effort that included outsourcing of non-core functions to the private sector.[30]

The UAE's political system can be characterized as authoritarian, tribal and federal. Political reform pressure has generally been low and major changes have not been notably advocated.[31] The most salient exception was a rather modest open petition, signed by 133 citizens in the spring of 2011, in the wake of the regional popular uprisings, calling for true legislative powers for the Federal National Council (FNC). Fearing further domestic demands, the government's counteraction to the Arab Spring took three forms: quasi-reform (amendments in the size of the FNC's electoral college), material bribery (announcement of new infrastructure projects, social spending for nationals, and federal government employee salary rises) and intimidation (imprisonment of five pro-reform bloggers and academics, among others).[32] The Emirati population is tied into a rentier bargain through a dense web of material welfare benefits (free education, subsidized natural resources, financial support, and land and home allocations, to mention a few) and immaterial legitimacy resources, based on kinship and proximity to the ruling elite. Nevertheless, direct and indirect welfare benefits are strongest in Abu Dhabi; hence future calls for reform are more likely to come from the smaller emirates. In terms of political reforms, the UAE continues to score poorly on both the Freedom House freedom index and the Economist Intelligence Unit's democracy index: in 2010, the country ranked 148th of 167, behind countries with a notorious reputation, such as Zimbabwe and Yemen.[33]

Seven times as complex

In addition to the tribe-based, inherited succession mechanism, it is the federal political system that makes the UAE unique among its neighbours.[34] Since its formation in 1971, there has been some evolution towards a federation, with key powers, including defence, foreign policy and immigration, having been ceded to the federal government. Nevertheless, the UAE is often described as a confederation, as centralization is still weak. Each member emirate still maintains a local government and retains complete sovereignty over its natural resources and fiscal policies.[35]

Because of its economic strength Abu Dhabi directly finances the federal budget: around 40% in 2010, with the rest coming from revenues earned by federal bodies.[36] It also holds *de facto* the permanent presidency and the federal capital, and controls foreign policy and the Union Defence Force. Since the early days of the federation, the emirate has abundantly financed the smaller emirates, and to some extent even the second wealthiest emirate, Dubai. In 2006–09, transfers to the four smallest emirates classified as domestic aid amounted on their own to US$2.6–4.4bn per year.[37] Abu Dhabi's influence over Dubai has always been a sensitive issue. On more than one occasion the former has rescued significant sectors of the latter's economy through federal intervention. This has considerably strengthened Abu Dhabi's influence over the second wealthiest emirate, particularly after the 2008 credit squeeze.[38]

Another marked feature of UAE dynamics is inter-emirate competition, some of which can be judged as unprofitable from both federal and local viewpoints.[39] In the 1970s, major infrastructure and social welfare investments were made, which led to partial duplication of expenditure. The several international airports and the competition between Dubai's airline Emirates and Abu Dhabi's Etihad are just two examples of the persisting tendency. Duplication of official bodies has also caused problems in the energy sector,[40] and the existence of multiple agencies in charge of environmental governance has led to deficiencies, even tensions, in policy coordination and making.

The old energy

Because of its abundant oil reserves, Abu Dhabi's main external energy challenges arguably relate to global demand for and pricing of fossil fuels in the medium and long term. Potential exaggerations in oil reserve size, their maturity, and uncertainties regarding production capacity investments are additional challenges. Of Abu Dhabi's 12 primary fields, 11 were discovered between 1958 and 1969, and one in 1975. The largest field in the UAE, Zakum (1963), is alone estimated to hold 66bn barrels of proven reserves. Enhanced oil recovery is increasingly applied and recent exploration efforts have been described as disappointing.[41] In order to maintain supply to international markets from its maturing oil fields and reap revenues for the growing population, important investments are needed in the coming years. Plans in 2010 were to expand the UAE's total production capacity from 2.7 million barrels per day (mb/d) in 2009 to 3.5 mb/d by 2019, including a 400,000 b/d capacity expansion by the Abu Dhabi Company for Onshore Oil Operations (ADCO), costing over US$4bn.[42] With these massive investments in the pipeline, sustained global oil demand and prices are crucially important for Abu Dhabi. According to the UAE President and ruler of Abu Dhabi, Sheikh Khalifa bin Zayed Al Nahyan, speaking in 2009, a fair price would have been US$70–75.[43] (Estimates of the country's break-even price in the same year ranged between US$40 and US$79.)[44] Most of Abu Dhabi's exports are currently destined for Asia (96% in 2009), where demand is expected to keep growing. Japan is the main crude oil export destination, with 369m bbl sold in 2008 and 286m bbl (US$17.6bn) in 2009.[45]

Abu Dhabi's fossil energy sector is organized under the major state-owned company, the Abu Dhabi National Oil Company (ADNOC, est. 1971), a conglomerate including fourteen subsidiaries specializing in different areas of the industry. The government continues to engage in joint ownership with foreign companies to secure top foreign expertise and technology. Arguably, this is in part also to safeguard the uninterrupted flow of oil to the world and back up the regime's continuity. Japanese, French, British and American oil companies own up to 40% of Abu Dhabi's fossil energy sector. The three main joint ventures, ADCO, ADMA and Zadco, control almost all oil production, accounting for 94% of the UAE's total production in 2008.[46]

Notwithstanding exploitation problems (discussed below), Abu Dhabi's proven natural gas reserves are relatively abundant: 5.6–6.0 tcm, constituting 3.2% of the global total.[47] The reserves are owned and managed by ADNOC and its foreign partners and are predominantly exploited for domestic purposes, including electricity and industrial use. In 2010, according to OPEC, 22.4 bcm were reinjected in oil reservoirs for enhanced recovery.[48] Abu Dhabi first considered developing its sour gas reserves, which constitute 50% of the UAE's total reserves, in the mid-1990s. A large part of the reserves are difficult to exploit and high extraction costs postponed implementation for over a decade.[49] The fact that most of currently exploited gas is associated gas, and hence influenced by OPEC production quotas, has also slowed down exploitation.[50] In the 2000s, a new emphasis was given to their development as a consequence of rising oil prices, growing electricity, water and industrial feedstock demand, and the need for enhanced oil recovery.[51] An ADNOC manager admitted in 2008 that local demand had gone beyond the planners' imagination and accelerated reserve development was urgent.[52]

This came partly too late, as testified by the gas balance sheet: in 1980–2008, the UAE's total final energy consumption grew over fivefold, and by 2008, domestic oil consumption had risen to 17% of total production.[53] During the 2000s, the UAE's domestic consumption of natural gas almost doubled, to 60.5 bcm in 2010. Production did not keep pace, reaching only 51.0 bcm that year. Abu Dhabi's LNG exports, which are mostly destined to Japan under long-term contracts, remained rather stable by comparison with previous years, at 7.9 bcm. However, in the same year the UAE had to import over double this amount (17.4 bcm) from Qatar. Over half of these imports went to Abu Dhabi's utility company ADWEC (Abu Dhabi Water and Electricity Company), which also provides electricity and water for the northern emirates and Sharjah.[54] Thanks to imports of generously priced Qatari pipeline gas from 2007 onwards, natural gas still constitutes the federation's main source of electricity: 98% in 2008.[55] However, mid-term prospects look bleaker. Despite eagerness at the receiving end—in Abu Dhabi, Dubai and Oman—and a total pipeline capacity of 90 million cubic metres per day (mcm/d), Qatar has not been willing to export more than 57 mcm/d, under the original agreements from 2003–05, because of price disagreements and the moratorium on Qatar's massive North Field. This has forced the UAE to seek alternative, mostly more expensive sources.

The extremely hot summers place an important stress on the UAE's power supply capacity: in 2011 summer peak demand in Abu Dhabi was around 60% higher (9 GW) than during the rest of the year.[56] A government-commissioned study from 2008 projected that the UAE's annual peak demand for electricity would rise by 9% per year from around 13 GW in 2007 to over 40 GW by 2020. Natural gas would be able to supply around half of the needed capacity, and renewables 7% at most.[57] Government estimates remained relatively high even despite the global economic crisis. In 2010, ADWEC affirmed that the global situation had relatively little impact on local (peak) demand growth, which it projected would triple from 6.3 GW (gross) in 2009 to 20 GW in 2020.[58]

To enhance domestic energy security and for source diversification, Abu Dhabi opted in 2008 for nuclear energy (discussed in a case study in chapter 5), which is expected to cover around 25% of the emirate's power needs by 2020 when four 1.4 GW nuclear reactors are planned to be operational. Later on, nuclear power could account for a similar share of the entire federation's electricity supply.[59] Meanwhile, the other emirates have proceeded with independent plans, arguably for two reasons: nuclear electricity will not be available until the late 2010s at the earliest, and there is still some uncertainty regarding whether, when and for what price it will be made available for the other emirates.

Abu Dhabi has also announced a 7% renewables capacity target by 2020, which, with the above mentioned ADWEC figures, would translate into 1.33 GW. The reasons for the sluggish deployment of renewables are the same as in the other Gulf monarchies: the low cost of fossil fuels in relation to the cost of solar energy; the high utility subsidies; and the absence of international obligations to curb greenhouse gas emissions, stemming from the GCC states' developing country status in the UNFCCC. The government has been planning to introduce feed-in-tariffs for renewables, and an energy policy document that would pave the way for subsidy mechanisms for renewable energy has been expected since 2009.[60]

A UAE government policy paper in 2008 excluded coal as an environmentally detrimental and supply-wise risky option, but this is presumably an Abu Dhabi perspective, given that the emirate produced the study. Coal still remains on the table in other emirates.[61] In 2011, without further elaboration, Dubai announced that 70% of its energy mix by 2030 would come from natural gas and the rest from 'clean coal' and nuclear and renewable energy. Dubai has also flashed modest renewable energy targets of 1% by 2020 and 5% by 2030, in what some experts regard as an under-promise-over-deliver strategy. Towards this goal, a US$3bn 'Mohammed bin Rashid Al Maktoum solar park', reaching a total capacity of 1 GW by 2030, was launched in 2012.[62]

The stickiness of consumer subsidies creates a heavy financial burden: in total US$18bn for the entire federation in 2010.[63] The UAE ranks among the top three per capita consumers of energy in the world, and the residential sector takes up roughly a third of all electricity consumption (40% in Abu Dhabi).[64] In 2010, producing a kWh of electricity in Abu Dhabi cost on average 27.6 fils (US¢7.5), while industry, the commercial sector and non-Emiratis paid 15 fils/kWh (US¢4.0) and Emiratis only 3–5 fils/kWh (US¢0.8–1.3).[65] Government subsidies to the water and electricity authority ADWEA have been rising phenomenally since their introduction (in the balance sheets) in 2004, and stood at AED9.7bn (US$2.6bn) in 2009.[66] The situation at the pump has been until recently similar: the average petrol price in 2006–08 was 52–80% of the average US retail price,[67] and according to an HSBC estimate, fuel subsidies per capita in 2009 stood at US$2,290.[68]

The economic unsustainability of this situation for the six oil-scarce emirates—and Abu Dhabi bankrolling these—has prompted the federal government (which means in fact Abu Dhabi) to initiate a gradual liberalization of petrol prices, with three price hikes in 2010 alone.[69] In the autumn of 2011, at US$0.47, petrol prices in the UAE were the highest in the GCC, but still too low for Dubai, which imports its petrol from the international market (unlike Abu Dhabi, which produces petrol from its own crude oil).[70] Estimated financial losses of US$735m for 2011 alone, incurred by the Dubai government-owned Emirates National Oil Company, prompted the company to issue an unusual public statement in October 2011, in which it declared that the existing pricing mechanism was not sustainable and called for 'support of the concerned authorities in addressing the concern'.[71]

Water desalination is also massively energy intensive. Abu Dhabi predominantly uses water produced in cogeneration plants that desalinate by using steam from electricity production as the energy source.[72] In 2009, Abu Dhabi alone produced 962 mcm of water, and announced plans to double this during the 2010s at a cost of US$20bn.[73] In addition to the pressure on the Gulf marine environment, gas-fuelled desalination is also an important source of greenhouse gas emissions. According to the Abu Dhabi's environment agency, the share of water production of generation-related emissions in cogeneration plants is between 20 and 45%, depending on the season.[74]

For long, the government's primary policy tools in the electricity sector were to increase supply and its cost-effectiveness. Since late 1990s, Abu Dhabi has privatized its electricity sector and has managed to attract foreign investment and savings in terms of cost per unit produced.[75] Major capacity expansion plans, aimed at electricity production and water desalination, were announced in the late 2000s by the two government-owned gas companies, Gasco (for onshore operations) and Adgas (for offshore).[76] ADNOC is also pushing ahead with its massive Shah sour gas project, which has suffered repeated delays: in 2010, ConocoPhillips pulled out from an US$10bn agreement it had signed with ADNOC for reaching production of 5.6 bcm/year by 2015, complaining of lack of profitability. The company was replaced in early 2011—on enhanced contractual terms, it was reported—by Occidental, also US-based.[77] In 2011 Abu Dhabi's Mubadala bought a US$23m stake in a Tanzanian oil and gas field, in what some analysts have interpreted as an attempt to secure LNG imports for domestic use in a longer term.[78]

While the 2008 financial crisis came to partially save the situation—as it cooled down economic growth, particularly in neighbouring Dubai—power demand growth soon regained its former speed. ADWEA's power demand has been rising rapidly, driven by increasing exports to the five northern emirates since 2006. Between 2008 and 2011, Abu Dhabi's own peak demand grew by a massive 11% per year on average, while exports to FEWA and SEWA, the corresponding authorities of the four smallest emirates and Sharjah, increased by 34%.[79]

During the 2010s, the domestic side of energy security will constitute a major concern for the UAE. Demand side management only appeared among the government's policy tools towards the late 2000s. Since the gas crunch the UAE, with Abu Dhabi at the lead, has gradually begun introducing energy and water efficiency measures and awareness campaigns. Abu Dhabi's Executive Affairs Authority has been preparing a comprehensive demand management strategy for electricity and water consumption in the emirate, but by 2012 none had yet been published.[80] In 2009, a government-commissioned study found that the equivalent of two power stations could be saved by awareness-raising among consumers and that efficiency measures in irrigation and households could reduce water consumption by 30%.[81] A Water Master Plan for the emirate was launched in 2009 by the Environment Agency—Abu Dhabi (EAD), which has also led a campaign to install water-saving devices in 100,000 homes and public buildings.[82] Some elite members have suggested that fiscal measures would be implemented in the near future.[83] But as the softness of the measures planned and implemented by 2011 indicated, the Abu Dhabi government was still able to afford to treat subsidy cuts as an uncrossable line.

Inter-emirate energy politics

The inter-emirate level adds another layer to the resource equation. The northern emirates' increasing energy shortages are a serious issue for the poorer emirates themselves, but also a burden for Abu Dhabi, as it is deeply tied in a costly federal-level rentier bargain with them. Especially the four smaller ones (Ajman, Ras al-Khaimah, Fujairah and Umm al-Quwain), and increasingly Sharjah too, depend on Abu Dhabi for both infrastructure development and key natural resources. Around 2007–08, lack of local supplies and limited feedstock availability from Abu Dhabi began causing structural power shortages, occasionally blackouts, which placed limitations on local industries and led to important delays in residential and commercial projects in Sharjah and the poorer northern emirates. Abu Dhabi and Dubai were saved, thanks to supplies from Qatar.[84]

Already in 2008, Abu Dhabi and the Federal Electricity and Water Authority (FEWA, est. 1999) signed an agreement under which the emirate is responsible for providing energy to the four smallest emirates, and Abu Dhabi has announced a national power distribution grid, to be operated under FEWA.[85] FEWA, however, only originally agreed to supply for an increment of 7% each year. Meanwhile, demand in some emirates rose threefold. Abu Dhabi's reluctance to continue supplying for its smaller neighbours' construction boom was also evident in the announcement by FEWA in early 2008 that it would only supply electricity for residential projects.[86] In late 2008 Ras al-Khaimah was reported to have around 2,000 new buildings lacking connection to the grid, and an entire port relying on generators. In Umm al-Quwain, a US$8bn megaproject was put on hold. Even in 2010, at least 900 houses and commercial buildings were reported to lack access to electricity in the northern emirates.[87]

Because of the unreliability of short-term supplies from Abu Dhabi, the other emirates devised a number of other plans, ranging from a gradual dismantling of government subsidies for non-nationals,[88] through gas imports from other countries, to exploring alternative sources of energy. Dubai, which only owns around 0.1 tcm (2% of federal total) of remaining gas reserves, has imported pipeline gas since 2007 from Qatar via Abu Dhabi and, since 2010, LNG also.[89] Sharjah, which owns some gas reserves (0.3 trillion cubic metres, 5% of the UAE's total), has similarly had to resort to imports from Abu Dhabi. Sharjah has for almost a decade sought Iranian gas, so far unsuccessfully.[90]

In the 2000s boom period diesel generators, despite being significantly more expensive and polluting than natural gas, became a common temporary solution for both individuals and businesses in the smaller emirates. The smaller emirates also sought a diverse range of more permanent solutions, including joint oil development projects and alternative sources. Ajman and Sharjah agreed to develop jointly the offshore Zora gas field, and Ajman, Ras al-Khaimah, Fujairah and even Dubai, signalling their desperation, announced in 2008–09 that they were considering coal as an energy option.[91] Ajman was the first to sign a US$2bn deal in 2008 with a Malaysian power producer for a 1 GW plant that would be operating by 2012. Two Abu Dhabi-financed electricity plants were to be built to supply the needed energy meanwhile. However, the coal plant was later postponed until further notice.[92] Similarly, in 2009, Ras al-Khaimah announced it would start importing coal from Indonesia for a new 600MW plant to be built in the emirate, with the possibility of later expanding this to 3,000 MW.[93] Fujairah too has studied the viability of coal power. However, because of the problems associated with coal-fired electricity—reliance on imports, large storage facilities, and high carbon dioxide emissions—it might not be a realistic option,[94] particularly if and when cheap nuclear electricity from Abu Dhabi becomes available.

The smaller emirates have also given some consideration to renewable energy, though mostly on a small scale. Examples include floating 'solar islands', developed and tested in Ras al-Khaimah since 2007.[95] Following Abu Dhabi's example, the other emirates have also sought to increase private sector participation in their water and power sectors as a partial solution to existing inefficiencies. However, despite a 2008 law that allowed private power plants, the high utility subsidies continue to largely repel private interest in the sector.[96]

In 2011, chronic power shortages still continued to plague the five northern emirates.[97] But unexpected events, in the form of revolutions and popular uprisings in Tunisia, Egypt, Bahrain and elsewhere, quickly changed this. Fearing open expressions of discontent by less wealthy Emiratis, the Abu Dhabi government made a series of promises to improve welfare in the northern emirates, and that of Emiratis more generally. The promises included an announcement in March 2011 of a US$1.55bn expansion in the northern emirates' power and water electricity sector.[98] The Ministry of Economy also launched a stabilization campaign for 400 major food commodities.[99] As a short-term, albeit temporary and expensive solution as Mills has noted, Abu Dhabi also announced, in the spring of 2012, the construction of an LNG import terminal in Fujairah, on the Gulf of Oman. Once operational, the terminal could cover up to a fifth of the federation's consumption.[100] Most important, as elsewhere in the GCC, the Arab Spring had a devastating effect on the utility subsidy reform that had only just started: concerned about popular sentiments, in a highly popular move among local industries and businesses Dubai announced in August 2011 a freeze in tariffs for the 'next few years'.[101] With a total population of several million, the stalling of badly needed energy price reforms for populist motives will undoubtedly place a heavy cost on the national budgets. The consequent rise in the country's break-even price for oil will also render the UAE, like other GCC governments, increasingly dependent on high oil prices.[102]

Human agency: the environmental legacy of Sheikh Zayed

The significance of Sheikh Zayed in creating what can be termed a fed-eral 'culture of patriarchal environmentalism' is crucial for understanding the contemporary environmental sustainability discourse of the Abu Dhabi government. Sheikh Zayed bin Sultan Al Nahyan, born around 1917–18, has been described as one the most distinguished political leaders in the Arab world, and 'a singular figure of immense charisma'.[103] Davidson argues that Zayed's 'personal vision and energy... provided the polity with its ultimate patriarchal figure'. Zayed ruled the emirate of Abu Dhabi from 1966, and the UAE from its independence in 1971 until his death in 2004. He first gained influence as the governor, *wali*, of the town of Al-Ain, which he transformed into both a regional economic hub and the 'garden of the Gulf'.[104] Among Zayed's achievements in Al-Ain in 1946–66 were the restoration and maintenance of the local irrigation (*falaj*) sys-tem and provision of free water to small landowners to encourage local agriculture. He also had a defining role in the physical and psychological development of the 'Emirati nation'—hence the title Father of the United Arab Emirates—and enjoyed wide and strong support of the citizens of the UAE. Reinforcing the value of the nation's past—by promoting 'her-itage' though different material and immaterial ways, such as traditional sports and poetry and care for the environment—became a part of Zayed's legacy.[105] Intergenerational justice, preserving the land for future genera-tions, also figured prominently among Zayed's values.[106]

Nationally regarded as a conservationist, Sheikh Zayed, who 'greened the desert', also earned the title Man of the Arab Environment. His tra-ditional-style 'enlightened environmentalism' was based on planting trees and preserving wildlife.[107] Caring for nature was also often invoked by Zayed as an Islamic duty, as part of the 'triad of modernity, Islam and tradition' of power legitimization, common among the GCC states.[108] In order to fight desertification, Sheikh Zayed is estimated to have man-dated the planting of as many as 200 million trees. Among other things, he regulated hunting of endangered animals (in 1976 in Abu Dhabi, in 1983 in the UAE), founded a zoo in Al Ain (1967), and established a wildlife sanctuary on the island of Sir Bani Yas (1971).[109] Sheikh Zayed's conservationism also brought international recognition, most saliently in 2005, when the UN Environment Programme posthumously granted him the Champion of the Earth award.

After his passing, Sheikh Zayed was peacefully succeeded by his eldest son Sheikh Khalifa bin Zayed Al Nahyan as Emir of Abu Dhabi and President of the federation. Sheikh Khalifa, in turn, handed the titles of Crown Prince, chairman of the Executive Council and deputy supreme commander of the UAE Armed Forces to his younger half-brother Sheikh Mohammed. Sheikh Khalifa, whose succession took place by unanimous decision of the Supreme Council of Rulers, is a less visible figure, but continues to enjoy legitimacy among the local ruling families and general population.[110]

While Sheikh Zayed had several wives, sons and daughters, the so-called Bani Fatima, comprising the six sons of his favourite wife Fatima, have since the mid-2000s been turning into an increasingly important group within the ruling family, controlling many of the emirate's key portfolios. The most powerful figure is the group's senior member, the dynamic Crown Prince Sheikh Mohammed, with his youngest brother Abdullah, currently Foreign Minister, as a strong partner and supporter. The others (with important posts in brackets) are Sheikh Hamdan (the Ruler's representative to the Western Region, chairman of the Environment Agency—Abu Dhabi), Sheikh Hazza (national security adviser), Sheikh Mansour (Minister of Presidential Affairs and Deputy Prime Minister since 2009), and Sheikh Tahnoon.[111]

The emirate's accelerating pace of reform and economic development is generally attributed to Mohammed's growing influence over decision-making.[112] Since 2002, he has been the chairman of the Mubadala Development Company, the main investment vehicle of the Abu Dhabi government. Consequently, the Masdar Initiative (analysed in chapter 5) is also under Sheikh Mohammed's directive umbrella. He is also head of the UAE Offset Program Bureau, the Urban Planning Council, and the Abu Dhabi Education Council, among others.[113] In addition to Sheikh Mohammed, another key person in Abu Dhabi's non-oil economy and its domestic decision-making is Khaldoon Khalifa al-Mubarak, who holds the posts of chairman of the Executive Affairs Authority and the Emirates Nuclear Energy Corporation, vice chairman of the Abu Dhabi Urban Planning Council and CEO of Mubadala.[114]

Institutions and decision-making dynamics: the broader picture

The United Arab Emirates' government is officially both a federal presidential system and a constitutional monarchy. The Supreme Council of

Rulers, formed by the seven emirs, acts as the ultimate executive and leg-islative power, and elects the President (by convention, the ruler of Abu Dhabi), the members of the Council of Ministers and the judges of the Federal Supreme Court. Each emirate has a minimum of one post in the Council of Ministers, which is led by the Prime Minister (by convention, the ruler of Dubai, who is also the Vice President), but Abu Dhabi and Dubai hold the majority and most senior of posts.[115] In 2010, the coun-cil comprised twenty-four ministerial posts.[116] Political parties are pro-hibited and the 40-member legislature, the Federal National Council (FNC), enjoys only consultative status. In 2006, for the first time, half of its members were chosen through indirect elections. In 2011, the size of the electorate was increased to 129,000. In the absence of a true legisla-tive power, the Council of Ministers initiates legislation, which is then ratified by the Federal Supreme Council. In general, with the exception of the FNC elections, reforms of the political system have been practi-cally non-existent.[117]

The most important local level governments are the Executive Coun-cils of Abu Dhabi and Dubai, chaired by their respective Crown Princes. The Abu Dhabi Executive Council is the emirate's top decision making body. Alongside the newer institutions, the traditional institution of rul-er's *majlis* still exists.[118] The Abu Dhabi Executive Council chairman Sheikh Mohammed receives strategic analysis and policy advice from the Executive Affairs Authority. The Executive Council, which holds weekly meetings, consists of chairmen of the departments and some agen-cies, and members appointed by Emir Sheikh Khalifa. Abu Dhabi's gov-ernment is divided into five ministry-like departments. Additionally, there are seven councils, including the Urban Planning Council, and a number of autonomous agencies with specified powers, including the Environmental Agency-Abu Dhabi (EAD) and the Abu Dhabi Water and Electricity Authority (ADWEA), as well as regional and municipal authorities.[119] Some observers attribute the increased role and weight of Abu Dhabi's agencies in the 2000s to Sheikh Mohammed's personal pursuit of a stronger power base for his future role as the President of the federation.[120]

In energy policy-making, the sovereignty of the individual emirates over their natural resources leaves the federal Ministry of Energy (for-merly Petroleum and Mineral Resources), led by Mohamed Dhaen al-Hamli, with a largely ceremonial domestic role.[121] In the early 1990s, the

powers of the ministry were significantly cut down to policy coordination, meaning mainly OPEC representation and 'subsidies at the pump'.
Other tasks were taken over by the Abu Dhabi Supreme Petroleum Council (SPC), established in 1988 and headed since then by Sheikh Khalifa.[122]
Abu Dhabi's main hydrocarbons conglomerate, ADNOC, is governed
by the SPC, which therefore controls the emirate's oil policy.[123] Abu
Dhabi, owing to the size of its resources and its pre-independence membership (1967), is the only one of the seven emirates that participates in
OPEC's decision-making and is bound by its decisions.[124]

Reflecting a GCC-wide trend, long-term strategic development planning appeared on the UAE government agenda in the late 2000s. The
most important policy documents in Abu Dhabi were the Plan Abu
Dhabi 2030: Urban Structure Framework Plan, of 2007, and the Abu
Dhabi Economic Vision 2030, of 2008, both mandated by Crown Prince
Sheikh Mohammed. The documents lay out a grand master plan, which
aims at coordinated urban planning and a more diversified and sustainable economy. The Economic Vision defines four priority areas: economic
development; human resources; infrastructure development and environmental sustainability; and improving government. Among the main
objectives defined are reducing oil dependence and creating employment
and better education for the nationals. The government will focus a number of strategic economic sectors, in both the oil-reliant economy (such
as heavy industries) and the 'new', production-oriented economy (including tourism, aviation and aerospace and the health care industry).[125] Other
important documents include the medium-term Strategic Plan 2008–
2012 for the streamlining and increased coherence and coordination of
the different departments and agencies. As part of Sheikh Mohammed's
new approach to government, all government entities are required to
deliver updated five-year plans annually.[126]

Neopatrimonial power structures and dynamics in Abu Dhabi, in particular, create a number of challenges across the decision-making system.
Government officials are often appointed on consideration of social status, rather than competence, and personalities are said to be crucial in
determining the status of government departments.[127] Lack of local inter-
agency cooperation has been a major problem too, in the environmental
sector among others. According to a UAE-based stakeholder, the late
2000s saw visible improvement in this area, including improvement in
the flow of information and integration of plans and actions. Interaction

between peer departments and agencies in the different emirates, and even between local and federal level institutions, can still remain very limited, however.[128]

Abu Dhabi's environmental and climate governance

Because climate change is an environmental problem with an energy solution, its governance lies at the crossroads of energy and environmental institutions. In the case of Abu Dhabi, the traditional energy industry and institutions have largely stayed clear of domestic climate change governance, leaving space for two emerging organizations: the government-owned company Masdar and the Environment Agency—Abu Dhabi.

Abu Dhabi's environmental and climate change governance is dominated by a handful of influential figures and institutions, with Crown Prince Sheikh Mohammed in the lead. As the chairman of the Abu Dhabi Executive Council, he is the motor of Abu Dhabi's new economy, to which alternative energy and sustainability-related initiatives belong. Local stakeholders describe environmental issues as one of the three priorities on Sheikh Mohammed's agenda, the others being education and security (with the domestic dimension of this latter one receiving increasing emphasis since 2011). According to some interpretations, Mohammed's interest in the environment is inherited from his father, Zayed.[129] Sheikh Abdullah, the Foreign Minister, is responsible for the UAE's nuclear policy vis-à-vis external audiences. He has since 2009 also been involved in the renewable energy and climate change-related endeavours of the emirate through the International Renewable Energy Agency headquarters and the ministry's Directorate on Energy and Climate Change. Sultan al-Jaber, an Emirati technocrat, is the founding CEO of Masdar, which arguably planted the seed for most subsequent climate change-related policy developments in Abu Dhabi. He has since 2010 also served as the UAE's special envoy for energy and climate change, vice president of the National Climate Change Committee, and for a while as the UNFCCC lead negotiator. In the past years, Abu Dhabi's environmental institution, the Environment Agency—Abu Dhabi (EAD), until recently led by Majid al-Mansouri, has been granted an increasingly important and visible role in local environmental governance. As a newcomer, the long-term managing director of the UAE's most important environmental NGO, the Emirates Wildlife Society, Razan Khalifa al-Mubarak (Khaldoon al-Mubarak's sister), was in 2011 appointed the EAD's secretary general. Also, the federal Ministry of Environment and Water, currently led by Rashid Ahmed bin Fahad, a 'technocrat environmentalist' from Dubai,[130] nominally guides environmental and climate-policy-making in Abu Dhabi.

From a historical perspective, environmental considerations first appeared on the UAE government's agenda in the early 1970s when municipalities were mandated to preserve the environment. In 1975, the Higher Environment Committee was established under the Council of Planning to 'link environmental considerations to planning and development policy'. Ensuing from the UAE's participation in the Rio Earth Summit of 1992, the Federal Environment Agency (FEA) was established under the Ministry of Health in 1993, replacing the Committee.[131] While the FEA was responsible for developing legislation and environmental standards, the local environmental authorities (in Abu Dhabi, Sharjah and Ras al-Khaimah) or municipalities (in the other emirates) have held the responsibility for implementation of the federal environmental laws and preservation.[132] The Ministry of Environment and Water was established in 2006, reflecting a growing emphasis in the area.[133] Prior to FEA's full incorporation into the Ministry, the two coexisted until 2009.[134] Today, the Ministry is still a weak institution—in part, presumably, because of its short existence, but also because of a lack of funding and capacity.[135] The EAD's strength and independence have probably also played a role.

The EAD is the most capable and best-resourced local environmental authority, not only in the UAE but in the GCC. Established in 2005, it replaced the Environmental Research and Wildlife Development Agency (ERWDA, est. 1996) that had been funded by the patronage of Sheikh Zayed, who had also established in the late 1970s the National Avian Research Centre.[136] FEA's and ERWDA's relationship up to the early 2000s was characterized by a lack of cooperation and fierce competition 'for both funding and recognition as a driving force of environmental issues'. Originally a research-oriented institution, the EAD's role was quickly expanded towards that of a competent local authority.[137] Currently the agency's scope includes environmental management and policy, biodiversity management, and environmental awareness-raising.[138] It has also received increasing visibility through sponsoring and organizing a large number of international and regional environmental and sustainability conferences.[139] Partly validating the assumption that the relative importance of a bureaucracy in the UAE is reflected in the size of its budget and human resources,[140] the EAD currently has over 500 employees. It has been lauded for excellence at the local level and admired as the spearhead of strategic environmental planning in the Arab world.[141] In 2009, former secretary general al-Mansouri won the GCC award for Best Environmental Personality in the UAE.[142]

As stakeholders have noted, the UAE exhibits a culture of establishing committees to persuade or even 'force' the different institutions to work together. If an issue is 'important', higher committees are formed from the top executives of each relevant entity.[143] As the issue of environmental sustainability policy-making began climbing on the policy agenda, the late 2000s saw a large number of new committees, including four UN climate convention-related committees and numerous emirate-level committees, such as Abu Dhabi's energy demand side management committee and permanent committee for water and agriculture.[144] However, this committee culture, grounded in the (neo)traditionalist ideal of consensus-based decision-making, often does not help to prevent inter-institutional competition and disequilibria, as shown further below.[145] And, as noted above, committees in the Gulf are also notorious for their lethargic impact on decision-making.

The years 2007–10 marked a turning point in Abu Dhabi's environmental governance and policy-making. Although locally based experts noted that federal-level coordination, knowledge and know-how were still lacking, they also perceived a clear increase in government interest in climate change-related issues.[146] Around this time, a new dynamism began to emerge, owing to the participation of new actors, including Masdar, the Urban Planning Council (UPC), and the Ministry of Foreign Affairs. This momentum was accompanied by the increasing activeness and professionalism of the EAD, which not incidentally moved, in 2010, into the same premises as Mubadala (owner of Masdar), the UPC, and the Executive Affairs Authority, all under the umbrella of Sheikh Mohammed bin Zayed.

Environmental regulation, planning and policies: theory and practice

Owing to the previously marginal importance given to environmental issues in the UAE, and the consequently uncoordinated approach, environmental authorities were established late and had weak relative weight in federal and local-level decision-making. For the same reason only marginal attention was given to regulatory mechanisms, enforcement of legislation, and sustained planning. Until the very late 2000s, a command and control approach prevailed in the UAE's environmental policy over economic instruments and awareness raising, as observed by Raouf. He has, however, praised the UAE's environmental legislation as the most developed of the GCC member states, noting that the government has actively sought to bring both local and federal legislation up to international standards.[147] Most of the UAE's environmental legislation was originally initiated by the UNDP and came into force only during the late 1990s and 2000s.[148] The core legislation comprises the law on the protection of the marine environment (law 23/1999), the law on the preservation of the environment (24/1999) and the law on wildlife trade (11/2002).[149]

Owing to the perceived 'development imperative', the development of oil resources was for long (and still is) seen as a priority overriding environmental issues. In the federation's first decades, this attitude resulted in important marine-based environmental degradation. Although legislation exists, its implementation and enforcement have been dubious. An obvious example is the significant damage to the marine environment from development activities. Dubai's land reclamation projects since the 1990s, for example, have been described in a UN study as 'particularly destructive' and accompanied by 'minimal environmental management attempted to mitigate negative impacts'.[150] Industries and real estate developments have also been overlooked in Abu Dhabi, according to some reports from 2007, which note that none of the major diversification projects to that date had been supported by mitigative environmental measures. These included an aluminium smelter, an airport expansion, two new ports and industrial zones, and tourism industry-related infrastructure and real estate projects.[151]

Also, because environmental laws exist at both local and federal levels, duplication of efforts and resources is often a problem. In the early 2000s, with the new federal legislation in place, some saw a 'clear move toward consistency across the federation'.[152] Still, in 2007, O'Brien *et al.* observed that there was not a clear organizational infrastructure and division of labour for instituting, monitoring and enforcing environmental legislation. Even the FEA, although entrusted with more influence and power by legislation, was reported to have lacked both human and financial resources, which rendered its impact and its ability to carry out initiatives marginal.[153]

A major grey area in environmental monitoring and enforcement exists even today in the area of the energy industry, owing to the national oil company's special status in the emirate's decision-making structures. ADNOC has its own sustainability initiative, energy efficiency efforts, marine environment protection programmes, and environmental codes, which according to an SPC adviser are even stricter than some international regulations.[154] In 2010, the company began publishing annual sustainability reports, which include a section on environmental performance. From the outside, however, ADNOC is perceived as 'a kingdom in itself' that plays by its own rules. Local environmental officials have implied that the EAD has no power over the oil giant in the sphere of environmental issues. Simultaneously, the environmental agency feels it needs to seek to engage the oil company because of the latter's massive environmental impact.[155] As an illustration of the implications of ADNOC's role for policy coordination, in 2008 the SPC refused to participate in the emirate's Environment, Health and Safety Management System (EHSMS) on the grounds that 'the ADNOC group of companies implements the EHSMS according to international standards and reports directly to the Executive Council'.[156]

As stakeholders have observed, the rising domestic energy security concerns in 2006–07 also elevated environmental sustainability issues to a new level in Abu Dhabi. The government, led by Crown Prince Mohammed, began pushing sustainability and environmental considerations into all policy sectors, and an obligation was placed on competent authorities to report to him on performance in these areas.[157] Despite their state-of-the-art design, many of the new planning and policy frameworks, discussed further below, are still at pilot or even drafting stage. Hence the success of their implementation remains to be evaluated in the years and decades to come. It is generally thought that Abu Dhabi has both the funds and the advantage of a previously slower pace of development than Dubai to 'get it right' in sustainable development.[158]

In the area of strategic planning, a national environmental strategy and action plan was prepared in 1997–2000 by the FEA and local UN agencies, and a strategy for combating desertification was approved in 2003.[159] By the late 2000s, federal-level planning had all but disappeared:

the Ministry of Environment and Water did not publish a single (English-language) environmental plan or strategy during its first years. The opposite has happened at the local level in Abu Dhabi. In the 2000s, the EAD launched two five-year emirate-level environmental strategies, for 2003–07 and 2008–12. The current Environment Strategy for the emirate outlines thirteen strategic priorities, including water resources, air quality, development of a climate change framework and a waste management policy, biodiversity, environmental awareness, organizational efficiency, and better communication.[160] In 2002 the EAD established the Abu Dhabi Global Environmental Data Initiative (AGEDI), aimed at improving the quality of domestic environmental data for decision-making purposes and enhancing coordination between the main public stakeholders. AGEDI also published an emirate-specific State of the Environment report in 2006 and was actively involved in the Rio Earth Summit of 2012.[161] Finally, in 2009, a team at the EAD began elaborating a long-term strategy under the title 'Abu Dhabi Environment Vision 2030'. The vision, which was soft-launched in mid-2012, will include a series of forecasts and scenarios, over five priority areas: water, climate change, air quality, waste and biodiversity. The Vision will form a broad framework for policy, governance, enforcement and regulation, and serve as the basis for five-year action plans.[162]

Around 2007, with sustainability emerging as a global trend in architecture and urban planning, both Dubai and Abu Dhabi also began incorporating sustainability aspects into their building codes and urban planning. Abu Dhabi's Masdar City, announced in 2007, is the UAE's first large-scale attempt at the full-scale incorporation of sustainable building. Despite its innovativeness, without emirate-wide regulation Masdar's direct sustainability impact would, nevertheless, have risked remaining limited to the walled community. In 2007 the Urban Planning Council (UPC) published a visionary urban framework plan, Plan Abu Dhabi 2030. The conceptual and physical limits of the previous plan, from the late 1980s, had been reached, and the population of the city of Abu Dhabi was expected to rise to over 3 million in 2030 as a result of the gradual relaxation of land ownership laws and increasing public investment in tourism, real estate and other areas of the new economy. Reflecting a lesson learned from Dubai's exponential growth, and lack of sustained long-term planning, the document stresses the need for coordinated growth and is grounded on three components of sustainability.

Sustainable resource use (efficiency, lower use of non-renewables and active exploration of renewables) is also among the objectives.[163]

A further step in urban sustainability planning was Estidama ('sustainability' in Arabic), a framework that concentrates specifically on buildings and communities. Originally an EAD initiative, Estidama was taken over by the UPC in 2008.[164] With important backing from the government,[165] Estidama has developed its own green building rating system, essentially an adaptation of the American LEED system, based on 'pearls'. It was incorporated into the UPC's planning approval and permitting processes in September 2010, after which all new projects must have achieved at least one pearl and government-funded buildings two pearls.[166] A year later, over 100 developments were already reported to have received this latter rating.[167] As a planner involved in process has remarked, Estidama feeds directly into the government's economic and energy diversification and emission reduction plans.[168] However, in a country where utilities are highly subsidized, government regulation will remain key for implementation, as economic advantages in sustainable buildings are mainly achieved through energy and water efficiency. Masdar City functioned in many ways as a pilot for the code's development and a feasibility showpiece, according to one of its directors, as local awareness surrounding sustainable construction had previously been low, even to the point of considerable resistance.[169] As critics have noted, with the new emphasis on long-term planning there is a danger that the government will overlook some of the pressing current environmental problems.[170]

Climate policy decision-making structures

At the emirate level, Abu Dhabi's Masdar and EAD have been the pioneers in pushing environmental sustainability and climate change-related considerations onto the local-level political agenda. Masdar in particular has been active in influencing certain national policy positions in the international climate negotiations since 2008.[171] In the energy sector, Masdar has had a seminal impact on the emerging regulatory framework through its projects in renewable energy (requiring feed-in-tariffs), energy efficiency of buildings, waste-to-energy[172] and environmental sustainability in general.

Among the institutions that have participated in the international climate negotiations, there are said to have been big egos and competition. One culmination point was reached in 2010 when the leadership of the

UAE's UNFCCC delegation (or the title of the National Focal Point; see also chapter 8) was transferred from the Ministry of Energy, which had held it since 1992, to the Ministry of Foreign Affairs and Ministry of Environment and Water jointly. Confusion over policy leadership that year was revealed when for a moment it was not clear which minister (Rashid bin Fahad and Sheikh Abdullah, both present) would lead the delegation in the UN climate change conference in Mexico.[173]

In 2009, inspired by the UAE's victory over the headquarters of the International Renewable Energy Agency (see chapter 8), according to close observers the Foreign Minister Sheikh Abdullah wanted to become involved in the Copenhagen UN climate conference (which was the highest profile climate change conference to date) and the ensuing high-level political agreement. That was how the Foreign Ministry-based Directorate of Energy and Climate Change (DECC) became established, in 2010. The Foreign Ministry also envisioned that its non-sector-specific mandate—and most likely its political clout also—would enable the DECC to appear more neutral and exert more power over the different sectors and emirates.[174]

Despite the Ministry of Environment officially being in charge of the country's external climate policy formation, the dynamic DECC quickly took a leading role. The DECC officially functions as the National Climate Change Committee's secretariat and is headed by Masdar's CEO Sultan al-Jaber. The directorate is modelled on similar departments in the UK and Denmark, but does not include an energy department. Some have speculated that it might eventually evolve into a ministry. In 2012, the DECC was still a small body. While it has been steering the federation's external climate policy to a markedly more progressive direction, its impact on the UAE's domestic level remains uncertain: this is partly because of the prevalence of the 'committee culture' and difficulties in reconciling the substantial and institutional interests of the two main emirates. Perhaps even more important, the DECC's clout might remain limited because of the paramount importance of oil in the economy. As suggested by UAE-based stakeholders, climate change remains lower on the national agenda than might appear from the outside, and climate policy will go only as far as it does not conflict with the interests of ADNOC and other key economic interests.[175]

From bad press to awareness-raising: perceptions and policy-making

The case of the ecological footprint is an excellent example of how external image is a concern for the UAE's elite, and of the way related perceptions can initiate policy changes. The low performance in international environmental rankings, and consequently in the views of the media, has been perceived by the image-conscious leaderships in Dubai and Abu Dhabi as harmful to the country's otherwise positive international reputation. The well-known Environmental Performance Index of Yale and Columbia Universities ranks countries according to their performance in the areas of environmental public health and ecosystem vitality. The indexes of 2008 and 2010 ranked the federation at 112[th](/149) and 152[nd](/163). In 2010, the UAE also performed the worst in the Middle East.[176] An even more widely recognized comparison, however, is the WWF's Ecological Footprint Index, which measures stress per capita on the use of natural and ecological resources within a country's borders (see also chapter 5). Since the beginning of the footprint calculations in 2000, until 2012, when Qatar and Kuwait surpassed it, the UAE held the world's highest ecological footprint. According to the 2012 index, in 2008 the lifestyle of an average UAE inhabitant, if extended to all the world's population, would require five planet Earths to sustain it.[177] At the height of the Dubai boom, these low environmental rankings, which received some global media attention, were perceived as bringing undesired bad press for the emirate. They also ran fundamentally counter to the 'greening project' initiated by Sheikh Zayed, which, as expressed by Ouis, 'has been tremendously successful in creating a global image of being a modern, environment-friendly society', and is closely linked to political legitimization.[178] As expressed by a prominent Emirati environmentalist, it was at this point that the UAE's leaders realized the need to 'keep up with the changing times'. To conquer its place as a global business hub, the government needed to start addressing issues like sustainability and transparency.[179]

After the publication of the 2006 Living Planet Report, in 2007, the Ministry of Environment and Water (led at the time by Mohammed Saeed al-Kindi) and the local branch of WWF, the Emirates Wildlife Society (EWS/WWF), established a project titled Al Basama Al Beeiyah (Ecological Footprint). One of the project's objectives was to produce more accurate national statistics for recalculating the country's ecological footprint. Although the government initially claimed that the 2006 data for the UAE were incorrect,[180] the new data produced by Al Basama Al Beeiyah for the 2008 report did not change the country's ranking. Most interestingly, the project represented a unique teaming-up of the government with an international environmental NGO. Furthermore, as the government discovered that a large share of the footprint came from the domestic sector, the cooperation expanded: together with the EWS/WWF, the EAD envisaged Heroes of the UAE,[181] the broadest and most visible environmental sustainability awareness campaign in the GCC so far. EWS/WWF has since then been consulted on a number of occasions by government institutions.[182] As an unforeseen display of NGO impact on decision-making (or co-optation, or even nepotism, as critics might argue), the organization's long-term managing director, Razan al-Mubarak, was appointed as the EAD secretary general in 2011.

Until the very late 2000s, the general direction of environmental awareness-raising had been top-down, and the government mainly supported single issues, such as the UAE-wide ban on non-biodegradable plastic bags by 2013.[183] Another visible form of awareness-raising is the patrimony-exuding high-profile environmental awards: the Zayed Prize (US$1m)[184] and the Zayed Future Energy Prize (managed by Masdar, US$2.2m).[185] Public environmental awareness was for long very low and only a small share of the population has been engaged in NGOs' activities. The EAD's annual awareness and behaviour survey from 2009 indicated that awareness and 'positive behaviour' among Abu Dhabi residents had slightly increased from the previous year.[186] Also, it is generally acknowledged that the temporary residence status of the majority of the population decreases general interest in sustainable behaviour and consumption patterns. The formerly inexistent 'infrastructure for environmental sustainability', including recycling facilities and public transport, has also hindered sustainability efforts of even the most conscious residents. Only recently has Abu Dhabi begun implementing recycling facilities and building the required reprocessing infrastructure for waste.[187]

Particularly in the 1990s, the government and businesses were the main patrons of local ENGOs. In particular the oil industry, according to Ouis, has played a leading role in promoting Emirati environmentalism.[188] As part of a broader international trend, the private sector has become increasingly involved in environmental activities through corporate social responsibility policies and, to some extent, by following new business opportunities. A 2009 survey of 32 UAE-based companies, for example, displayed an increasing awareness of environmental issues and 'motivation for action'.[189] In all sectors, however, greenwash arguably abounds. Some construction companies, including TDIC and Aldar, have sought to build a greener image, but construction companies in general are not considered to be interested in environmental sustainability.[190] It is in this sector that the UAE's developmental priorities—economic sustainability, growth and diversification—are still most manifest and largely override environmental sustainability.

The UAE's few environmental NGOs are practically all co-opted by the government. Traditionally, they have concentrated on education and awareness-raising through one-issue campaigns (recycling, beach clean-ups, tree planting) and lectures, and have generally not sought to influence policy. ENGOs in Abu Dhabi include the Emirates Natural History Group (established in the early 1970s, attracting mainly Western expatriates), the Environment Friends Society (1991, Emiratis), the Commission of Environmental Research of the Environment Friends Society (1999),[191] and the EWS/WWF (2001, professional NGO, under the patronage of Sheikh Hamdan bin Zayed).[192] Of these, all except the EWS represent a traditional conservationist approach rather than broader, more modern thinking that incorporates issues like sustainable development and climate change. As a new phenomenon, in the late 2000s a few individuals, like the Green Sheikh of Ajman, have emerged as role models for young Emiratis.

5

ABU DHABI'S CLIMATE CHANGE
AND SUSTAINABILITY RESPONSES

In the late 2000s, a set of energy and environmental sustainability-related initiatives, established under the wing of Abu Dhabi's dynamic Crown Prince Sheikh Mohammed bin Zayed Al Nahyan and his younger brother Sheikh Abdullah, initiated a transformation of the emirate's domestic agenda with regard to natural sustainability. The initiatives also created new ways for the world and the Emiratis themselves to perceive the emirate's role in the region. While fully aware that fossil fuels will most likely remain the backbone of Abu Dhabi's economic security and domestic energy security for the coming decades, from 2006 onwards the emirate's leadership took two significant strategic decisions in the field of alternative energy. The first was the establishment of Masdar, a pioneering, multi-faceted 'future energy' company which, through a feedback loop with international media, inspired the government to further action. Abu Dhabi has announced a 7% renewable energy target by 2020, secured the headquarters of the new 'IEA of renewable energy', IRENA, and rethought its international climate policy. This chain of events also empowered local environmental institutions to initiate a number of ambitious campaigns and policies to change the existing consumption patterns of energy and water. Closely linked was the second strategic decision, namely Abu Dhabi's civilian nuclear programme. Officially announced in 2008 and promoted by key supplier states, the programme envisaged making a significant impact on the entire federation's energy security starting from the late 2010s. Indicating the complex regional context in which the GCC states find themselves and the dual nature of nuclear technology, Abu Dhabi sought to present the programme to closely observing international and regional audiences as a model of peaceful, transparent and safe nuclear development.

Together, these two initiatives reflect broader changes in elite perceptions of and interest in the interplay of the global energy economy and international politics of climate change. And as this chapter will demonstrate, Abu Dhabi's responses to climate change and (un)sustainability, at the turn of the 2010s, were produced by the interactions of underlying structures of rentierism with the agency of top decision-makers and the domestic institutional context. Other factors shaping these responses were the elite's dynamic mechanisms of neotraditional legitimization and a number of external factors, including perceived pressures and opportunities, regional competition, and support from key external allies.

Climate change impacts

Physically, as a country with a hot and arid climate and low-lying, highly populated coastal zones, the UAE is extremely vulnerable to climate change. A World Bank study from 2007 classified the UAE as the second most vulnerable country to sea-level rise in the MENA region.[1] In addition to sensitive ecological subsystems, 85% of the UAE's population and 90% of its infrastructure are located on the coasts and sensitive to temperature rises.[2]

As a pioneer in the Gulf, the government has published three country-specific studies on the potential negative impacts of climate change: two national communications (2007 and 2010) to the UNFCCC, produced by the Ministry of Energy, and a three-part study published by the EAD (2009).[3] The UN reports describe the country's coastal zones, water resources, dryland ecosystems, agricultural production, human settlements, public health, and energy infrastructure as 'highly sensitive to climatic changes'.[4] Average annual temperatures are expected to rise by 2.1–2.8°C by 2050 and 4.1–5.3°C by 2100. Sea-level rise, in the form of inundation, erosion and flooding, is considered as carrying significant risks for the UAE's coastal zone. Depending on the level (the second study accounted for 1–9 m), the UAE could lose anything in the range of 1–6% (1,555–5,000 km²) of land by 2100. This would move Abu Dhabi's shoreline southward by 25–30 km.[5] Moreover, together with the over-extraction of groundwater resources and the resulting saltwater intrusion, rising temperatures and potentially lower rainfall would further increase the need for desalinated water and significantly undermine agricultural self-sufficiency policies.[6]

The EAD study highlights the vulnerabilities of the UAE's coastal zones and Abu Dhabi's water resources and dryland ecosystems. It calls for adaptation planning and cost evaluations for the impacts of sea-level rise and seawater intrusion on coastal infrastructure-dependent sectors, like tourism.[7] The study points out that current water consumption patterns and rising demand are the most urgent of threats, given that mixed results from precipitation models make it difficult to determine future impacts of climate change. Recognizing the sustainability paradox of Sheikh Zayed's legacy, it also notes that although it bears an important cultural value, 'serious consideration will [be] needed in determining the costs and benefits of continuing to green the desert, given the challenges of climate change'.[8]

Studies that have estimated the economic impacts of international climate change response measures underline the country's dual vulnerability. Apart from infrastructure vulnerability, the two communications by the Ministry of Energy cite several previous studies that project revenue or welfare losses for the country, even with the implementation of Kyoto's flexibility mechanisms.[9] On the other hand, an EAD-commissioned impact study, which included six policy scenarios running up to 2025, concluded that there will be little change in international oil demand at carbon tax rates below US$90/bbl and that international climate policy will only have limited effects on oil markets in the coming decades. Even with US$180/bbl tax levels, the study projects demand levels of over 80% of the zero carbon tax scenario. This leaves the oil producers with a 'window of time in which to prepare for the deeper cuts that could come later'.[10]

Evidently, without the simultaneous domestic changes taking place in climate change-related perceptions of Abu Dhabi's key environmental elite members, and a shift in the locus of policy-making from the Ministry of Energy to institutions more genuinely worried about climate change, this kind of study would never have been published, as it undermines OPEC demands for compensation for economic losses in the international climate regime (see chapters 3 and 8).

Resource scarcities, related vulnerabilities and adaptation measures

As discussed above, power supply might become a challenge as early as the 2010s even for Abu Dhabi, if both demand and supply sides are not promptly addressed. While nuclear energy is promising to realistically deliver in the 2020s, Abu Dhabi can always burn oil for peak or emergency supply should natural gas not be available from domestic reserves or regional/international markets. More serious problems, however, could emerge in the longer term in the areas of water and food security in the absence of swift policy measures.

Except for hydrocarbons, Abu Dhabi is poor in other natural resources and has only modest amounts of other minerals. Rainfall is extremely scarce, on average 78 mm annually, and deserts constitute 93% of the total land area. The emirate's harsh climate and increasingly scarce water resources render its water and food security crucial issues to survival of the state and its inhabitants.[11] The UAE has no lakes or rivers, nor does the federation share groundwater resources with neighbouring countries. Hence, in contrast to the Levant, transboundary water disputes are not a risk. Estimates of the federation's renewable water resources in the early 2000s ranged between 50 and 400 m^3 per capita, but by 2009, owing to population growth, it had dropped to 22 m^3, which is considerably below the water scarcity limit of 1,000 m^3.[12] In 2008, groundwater (mostly non-renewable) supplied 71% of Abu Dhabi's total water demand. The shares of desalinated water and treated wastewater were 24% and 5%, respectively.[13]

The UAE's water use patterns are extremely unsustainable, approximately twenty-six times greater than the available renewable freshwater resources, and total demand is increasing fast owing to population and economic growth, irrigation, and bad management of the resource. Abu Dhabi currently has 25,000 farms and 220 areas defined as forests, the latter having dramatically increased since the 1990s. 'Greening the desert' and irrigation for agriculture, forestry and plantations consumes over 80% of total water used in Abu Dhabi.[14] Ironically, agriculture only accounts for around 1% of the federation's GDP.[15] Domestic water use too is extravagant: on average 550 litres per day per person, with villa dwellers using up to 1,760 litres per day, mostly for garden irrigation. The UAE's green areas and over a dozen golf courses are also major consumers of water. Abu Dhabi's Yas Links golf course alone uses 5–7 million litres/day.[16] Most worryingly, at current consumption rates Abu Dhabi's directly usable groundwater aquifers will be depleted in 20 years, and all resources in 55 years.[17]

Short-term water insecurity is also acute: Abu Dhabi depends on six major desalination plants for drinking water, and the federation's strategic water reserves are sufficient for only two days.[18] In 2010, Abu Dhabi awarded a US$430m contract to an Arab-Korean consortium to build the world's largest underground water reservoir for long-term storage of desalinated water. An equivalent amount of three months' emergency supply, 18 million m^3, will initially be pumped into the storage, which could later be expanded to one year's supply. Diminishing groundwater resources and fears over the security of the country's desalination infrastructure (from natural or oil disasters or military or terrorist attacks) have been reported among the motives, and some analysts have linked the plans to Crown Prince Mohammed being 'a strategist and a military man'.[19] A parallel development supporting this analysis is a massive US$4.2bn oil pipeline, with a total capacity of 1.8 million bbl/d, that was inaugurated in the summer of 2012 and links Abu Dhabi's onshore oil fields to the port of Fujairah on the Gulf of Oman, hence bypassing the strategically sensitive Strait of Hormuz.[20]

Over time, the UAE's dependence on desalination is set to grow and might become more complicated, as the discharges of hyper-saline water endanger the local marine environment and may, together with rising temperatures, further increase the cost of desalination.[21] According to the EAD's estimate, even with new desalination capacity Abu Dhabi's water demand will be three times more than available supplies by 2050.[22]

Until recently, water management in the UAE has been lax. As noted by a Federal National Council representative, wastage of water is a 'luxury that is making us poor'.[23] The sovereignty of individual emirates over their water resources complicates federal coordination, and a federal law to establish a federal water agency dating back to 1981 (law no. 21) has not been enforced.[24] Owing to agricultural demand, inadequate pricing, wasteful consumption, and population growth, domestic water demand has been rising fast in absolute terms: from 495 to 609 million m^3 between 2000 and 2009.[25] This problem was noted in the Abu Dhabi Water Resources Master Plan produced by the EAD in 2009, which called for environmental and water management reform and the establishment of an Abu Dhabi Water Council for strategic planning and development. Leakages are also a major problem in the arid country. Currently roughly

20% of the water, or more, is estimated to be lost throughout the distribution system.[26] Among implemented policy measures to increase resilience and water security have been new desalination and wastewater treatment plants and recharge dams (in the northern emirates), restoration of *falaj* (irrigation) systems, and study of salt-tolerant crops (because of the brackishness of groundwater).[27] Newer policy measures and plans include hectarage limits on agriculture, market subsidy cuts and efficiency improvements, and reducing leakages through a new plumbing code and a specific programme. Even rainmaking is used: 97 flights carrying silver iodide flew over the emirate in 2009.[28] There are also plans to increase the share of recycled water that is used (from 50–60% to 100%).[29] Demand side measures in Abu Dhabi have largely been soft and concentrated on awareness-raising programmes (in agriculture and the residential sector) and publicly funded water-saving technology (tap devices in buildings and new irrigation systems and equipment on farms), rather than using pricing instruments to curb consumption. Fortunately, implementation is beginning to scale up, with 500 farms included in a pilot programme as of 2011. The government's ambitious plan is to cut agricultural water use by 40% by 2013.[30] At the federal level, the Ministry of Environment and Water has launched a six-part domestic water security plan (2011) and encourages farmers to shift to hydroponics (soil-less agriculture), which uses significantly less water. By 2010, the ministry had also drafted a new law regulating water use, which would have strengthened the ministry's role in the oversight of the UAE's water resources, including the establishment of a national water committee.[31] However, by 2012, owing to lack of clarity on the allocation of responsibilities,[32] but also as an indication of the reluctance of Abu Dhabi and others to surrender any powers to the federal level, the law had yet not been approved.

The UAE's high dependency on agricultural imports (roughly 90%) contributes to inflation and renders the country vulnerable to market crises and disruptions. During the world food price crisis in 2008, the UAE's food imports peaked at US$3.8bn. The Economist Intelligence Unit has estimated that the country's import bill will rise to US$8.4bn by 2020.[33] Concerned about future supply security and equipped with the needed financial surplus, Abu Dhabi has engaged in foreign land leases and purchases. In 2008, the Abu Dhabi Fund for Development declared it would develop 30,000 hectares (ha) of farmland in Sudan as part of 'a broader strategy to secure food supplies' in Africa, Asia (Vietnam and Cambodia) and South America.[34] In 2008 alone, several land agreements were reported, including 400,000 ha of land rights in Sudan bought by an unknown UAE investor and 324,000 ha in Pakistan by Abraaj Capital from Dubai.[35] In 2011, the farmland leases were reported to continue, and Australia, other Arab states, Romania, Turkey and—perhaps surprisingly—China had emerged as new (potential) partners.[36] Nevertheless, information regarding actual implementation of these deals and crop yield on the leased lands was scarce, possibly indicating lack of significant development. Building sustainable partnerships at all levels also remains a key challenge. As a foretaste of the potential security issues associated with farming in foreign lands, in 2009 a province in Pakistan banned deals between UAE-based private investors and local farmers amid concerns over the latter group's rights.[37]

Although in the GCC there is always a trade-off between water security and food self-sufficiency, the UAE is interested in maintaining a certain level of domestic production. On the domestic side, a major soil survey by the EAD from 2009 found that in theory 200,000 ha, in addition to the 77,000 ha already cultivated, could be used for agriculture, although this increase is not likely to materialize owing to domestic water security considerations.[38] In 2010, Abu Dhabi announced the establishment of a government-owned trading house, Abu Dhabi Sources, for securing food supplies internationally. A food security strategy is also being prepared by the Abu Dhabi Food Control Authority. A six-month strategic food reserve of fourteen commodities has also been under preparation since the 2008 crisis, and there have been plans to build storage silos in Fujairah in the event of Iran blocking the Strait of Hormuz.[39]

Recent adaptation strategies considered in Abu Dhabi as a response to the threat of climate change-induced sea-level rise have included developing strategic information systems for coastal zones; 'win-win strategies' that would incorporate both adaptation and (economically) sustainable development, such as protected areas; and raising awareness among coastal developers and real estate entrepreneurs.[40] Currently, the UAE on average retains strong financial adaptation capability. Tools available include building flood barriers, moving urbanization inland, and reclaiming land from the sea. Abu Dhabi's urban plan already concentrates largely on developing the city inland. Even so, the publication of the EAD's vulnerability study from 2009 can be considered as a courageous act, since the psychological effects of impact scenarios on potential international investors might be repellent, particularly in the case of projects on reclaimed land. It demonstrates the difference in mentality between Abu Dhabi, which is seeking to portray a genuine concern over climate change, and land rent-dependent Dubai, where economic realities still dictate development priorities. Broadly speaking, the UAE still lacks cross-sectoral and ministerial cooperation for addressing the challenges of climate change.[41]

Contribution to climate change and mitigation actions and policies

The UAE's total estimated CO_2 emissions (138 Mt in 2007) are small on a global scale (0.5%), and the country has low historical responsibility (if measured by domestic emissions). However, growth of emissions has been fast in past decades, and per capita figures are the world's second highest (32 t in 2007, see also table 8 in chapter 3). Apart from international estimates, systematic, nationally produced and verified, up-to-date data includes only two outdated UNFCCC emission inventories (1994 and 2000), an unpublished study commissioned by Masdar (which estimated the UAE's total emissions at 112 Mt CO_2e in 2005),[42] and ADNOC's reports on its own emissions (ranging between 23–25 Mt CO_2e in 2006–2010).[43] It can also be estimated from available data that ADNOC produces roughly 18% of the UAE's and over half of Abu Dhabi's total emissions. The Al Basama Al Beeiyah project found that in 2006, 83% of the UAE's ecological footprint was constituted by CO_2 emissions. Of the footprint, households accounted for 57%; businesses and industry 30%; and the government 12%.[44] Having recognized the need to produce accurate, up to date data for policy-making, Abu Dhabi has started preparing the ground for an inventory for 2010, expected by 2012.[45]

Up to the mid-2000s, only a few concrete larger-scale mitigation actions were taken in the UAE. These came as a result of policies and actions in the energy and transport sectors, the most significant ones probably being the reduction of gas flaring by 78% in 1995–2007 and a pilot scheme including gas-run vehicles.[46] In 2005, Kazim reported that 'strong opposition to developing renewable sources of energy [had] been fading in recent years'.[47] Some small-scale solar energy projects were implemented, including 33 GSM base stations (total 600 kW), pay phones (around 29 kW), cell enhancers (around 9 kW), 46 aviation obstructing warning lights (5 kW), and an 850 kW wind turbine on the

Sir Bani Yas island of Abu Dhabi.[48] In 2007, spurred by Saudi initiative, the UAE pledged US$150m to a clean tech fund formed by the four GCC OPEC member states.[49] Also starting the same year, several emirates became increasingly interested in the Kyoto Clean Development Mechanism (CDM). Starting with a landfill gas project in Sharjah, followed by a number of proposals by Masdar, the UAE had by mid-2012 achieved the approval of the Executive Board of the CDM for five small and medium-size CDM projects. Fourteen more were at validation stage. The other GCC states only had at most one registered project and five in the pipeline. The total expected annual greenhouse gas reductions of the five registered UAE projects corresponded to 996 kt of CO_2e (the equivalent of around 31,000 'UAE residents' emissions'), but were worth US$8.0m in November 2011 prices, even when carbon (CER) prices were at a record low, and falling.[50]

The first record of a specific climate change-related strategy is from the UAE's first UNFCCC communication in 2006. It stressed the importance of economic development priorities and the country's developing country status (and hence lack of emission reduction obligations under the UNFCCC), and proposed a number of win-win measures to cut emissions while enhancing growth. Signalling low real interest, these were measures described as 'an effort to demonstrate solidarity with the international community'. Notably, no alternative energy plans were presented.[51]

During the five years starting from 2006, the situation changed radically, particularly in Abu Dhabi. The initial push came in the decision of Crown Prince Mohammed to make knowledge-intensive and future-oriented alternative energy technologies a key element of the emirate's diversification drive.[52] This decision was embodied in the establishment of the Masdar Initiative in the same year (discussed further below). A number of related, coalescing factors also weighed in, ranging from the increasing precariousness of the domestic energy situation to the emerging negative environmental impacts of the 2000s' fast growth, and the rise of climate change on the international agenda. Important changes began taking place in elite attitudes towards alternative energies, environmental sustainability and climate change.[53] Also, the UAE had become conscious of the image impact of its environmental performance, and its top leaders, including Sheikhs Mohammed and Abdullah bin Zayed in Abu Dhabi and Mohammad bin Rashid Al Maktoum in Dubai, perceived a need to take an open approach to the country's high per capita emissions.[54] Moreover, with new studies on the potential physical impacts of climate change on the country, the issue of the impacts of international climate change response measures declined in relative importance on the UAE's UNFCCC agenda.[55]

In Abu Dhabi, a need for a climate policy or framework emerged. The Executive Council, led by Sheikh Mohammed, wanted the emirate to take a pioneering role in climate change mitigation while simultaneously keeping actions voluntary. In 2008, it asked the EAD to draft a strategic action plan, which would include incentive-based instruments but not numeric targets. The message, according to stakeholder accounts, was 'don't push it' so as not to overload the government with commitments. The plan would comprise observation networks, mitigation, adaptation and capacity building, and cover at least the water, transport, electricity and waste management sectors.[56] The EAD presented a final draft to national and local stakeholders in May 2009.[57] The Abu Dhabi-level strategy, however, was never published, firstly because of the low expectations for the Copenhagen climate summit in December 2009 and, later because of the takeover of the federation's climate policy in early 2010 by the newly-established DECC in the Ministry of Foreign Affairs, which incorporated it into its work on a federal-level strategy. (Undoubtedly, the plan, however, provided input for Abu Dhabi's Environment Vision 2030.) The strategy, not yet published at the time of writing, is expected to include mitigation programmes and cover all sectors, including oil and gas.[58]

The 7% renewables goal, an Abu Dhabi target, announced in early 2009 and linked to the Masdar Initiative and its investment plans, was mentioned as a federal level target in ministerial announcements in 2009–10, but seems since to have been forgotten. Dubai announced in 2011 a 5% goal for the emirate by 2030. Abu Dhabi-based climate policy experts asserted in 2010 that the UAE would most likely not take up any economy-wide emission targets, let alone commit to these internationally, owing to the difficulty of predicting future growth. They believe, however, that the UAE will not use its developing country status in the UNFCCC as an excuse not to do anything.[59] Many experts have criticized Abu Dhabi's renewables target as too easily attainable relative to the available time and financial resources, but local authorities have defended it as the first target from the Gulf and as a show of goodwill from an oil giant, pointing towards the economic unfeasibility of larger scale solar plants in the region at this stage of technology development.[60]

A change in rhetoric also took place. In 2009, for the first time, climate change mitigation was mentioned in the official *UAE Yearbook*, which stated that the country 'has a significant role to play in seeking ways to mitigate the impact of fossil fuels on our planet'.[61] The UNFCCC communication from 2010 noted that climate change demanded urgent and decisive action, which was a 'moral obligation to our children and their progeny' and that the UAE 'as an oil-exporting country, [had] already begun [its] journey towards sustainable development by introducing new thinking, new frameworks, and new partnerships for reducing [its] carbon footprint'.[62] In 2010, the EAD secretary general al-Mansouri listed water supply, pollution, and climate change as Abu Dhabi's priority concerns.[63] Finally, in early 2012, Sheikh Mohammed bin Rashid, Dubai's ruler and the UAE's Prime Minister, announced the launch of a ten-year National Green Growth Plan for the federation, which will include programmes and policies in energy, agriculture, investment, sustainable transport and construction and the environment.[64]

In 2009–10, as institutional awareness on climate change increased, officials of the Ministry of Environment and the EAD were able to name a number of domestic initiatives and policies with mitigation outcomes (but which had not necessarily been initially devised for this purpose). Among these were using natural gas for electricity generation; aiming to have 25% of Abu Dhabi's government vehicle fleet running on compressed natural gas (CNG) by 2012, and some of Dubai's *abras* (water taxis) also running on CNG; the zero flaring goal of Abu Dhabi's gas industry; Dubai's metro; a surface transport master plan in Abu Dhabi; a federal green building code and a planned transport policy; the nuclear energy programme; Masdar City with its renewable energy projects; and a domestic energy policy for Abu Dhabi (in the works), with demand side measures.[65] In August 2010, the UAE's key domestic mitigation initiatives were outlined in a UNFCCC conference as a pilot project in CCS, nuclear energy, renewable energy, energy efficiency (green building regulations and appliance standards), Masdar City, public transport infrastructure, and education and awareness programmes.[66] A number of individual sustainable building projects were announced starting from around 2008, including the Abu Dhabi World Trade Centre, Masdar Initiative's headquarters, the Shams Tower at the Yas Island F1 race track, and the developer TDIC's headquarters. By 2011, ten government build-ings in Abu Dhabi had been equipped with solar panels with a total capacity of 2.3 MW.[67] On the demand management side, many emirates (except Abu Dhabi) raised electricity and water prices to curb consump-tion. Also, the Heroes of the UAE campaign above described, targeted at consumers and businesses, aimed at changing the existing energy and water consumption patterns.[68] On the external side, the UAE provided development assistance to Pacific countries for climate change adapta-tion and to African states for renewables, and Masdar participated in a number of international joint ventures. Although good actions were many, their implementation was described as uncoordinated, and the big pic-ture remained largely undocumented by the competent authorities.[69]

On the planning side, in 2010, the Masdar Institute produced an elec-tricity and water scenario model for Abu Dhabi, which showed—with-out incorporating social or economic cost calculations—how the emirate could cut its CO_2 emissions by 11–38% by 2030 from business as usual. In a show of transparency, some of the results were made public in an Al Basama Al Beeiyah publication, even though similar calculations are politically very sensitive in the context of the UNFCCC, where coun-tries often seek to guard their achieved advantages and economic inter-ests. The scenario study showed that significant deviations from business as usual emissions, estimated at over 180 Mt in 2030 (compared to slightly over 40 Mt in 2005), could be achieved with a mix of planned develop-ments and energy and water supply demand policies.[70]

The case of Masdar[71]

We cherish our environment because it is an integral part of our country, our history and our heritage. On land and in the sea, our forefathers lived and survived in this environment. They were able to do so only because they recognized the need to conserve it, to take from it only what they needed to live, and to preserve it for succeeding generations.

Sheikh Zayed bin Sultan Al Nahyan, speech to mark the UAE's first Environment Day, 1998.[72]

Masdar was launched in April 2006, rooted in the principles of resource conservation and sustainable development practiced by the UAE's first President and Founding Father, Sheikh Zayed bin Sultan Al Nahyan.

Sultan al-Jaber, Masdar's CEO, 2010.[73]

A number of factors set Abu Dhabi's Masdar Initiative apart from similar natural sustainability initiatives and projects in the region's other monarchies, including its pioneering nature, multi-facetedness, scale, and specific focus on clean energy and environmental sustainability technologies. Despite the company's rocky path, Masdar's role as a catalyst to most of Abu Dhabi's renewable energy, climate change and environmental sustainability-related policy developments of the late 2000s is so crucial that it requires not only an examination of its first years of existence, including successes and problems, but also a deeper analysis of its functions in the vision of Abu Dhabi's green elite.

Working on the original plan, 2006 onwards

According to close sources, it was Sultan al-Jaber and three Lebanese consultants (Khaled Awad, Ziad Tassabehji and Osama Nader) who originally presented the idea of Masdar to Mubadala.[74] The Masdar Initiative, later known simply as Masdar, is owned by the private joint stock company Abu Dhabi Future Energy Company (ADFEC), which in turn is fully under Mubadala, set up by Sheikh Mohammed bin Zayed in 2002. Also dubbed as Sheikh Mohammed's investment vehicle, Mubadala's purpose is to advance economic diversification in the emirate of Abu Dhabi through a diverse portfolio in sectors such as energy, aerospace, real estate, technology, infrastructure and services. In the company's discourse, references are often made to the initiative as an extension of Sheikh Zayed's legacy.

Initiated in 2006, Masdar (Arabic for 'source') comprises a set of projects in the fields of alternative energy and carbon management, including research and development. According to the original plans, it would have culminated in a 'totally green city' of 40,000 residents and 1,500 businesses in 2016. Its four main aims are to diversify the economy, maintain and expand Abu Dhabi's position in global energy markets, transform the emirate into a developer and exporter of energy technology, and contribute to sustainable development.[75] Masdar comprises five units: City, Institute, Power, Capital and Carbon. Signalling the government's backing since its early stages, CEO Sultan Ahmed al-Jaber, Masdar's most visible and influential figure, was quoted in 2007 as informally saying that the initiative had 'an unlimited budget for renewable energy projects'.[76] As it would turn out later, this was not quite the case. As a promising start, in January 2008, the Abu Dhabi government announced a US$15bn investment in the initiative.[77] Of the five units, Masdar City, the CCS operations of the carbon unit, and the photovoltaics production unit were to receive the most funding. By mid-2009, Masdar had built up a portfolio including companies and projects over the entire technology life cycle, and had already invested US$3bn in its alternative energy and sustainability projects domestically and abroad, with an aim to reach US$10bn by 2015.[78]

The most visible aspect of the initiative, Masdar City, is a free zone that ranges over an area of 6 km^2 next to the Abu Dhabi international airport. Launched officially in May 2007, with a virtual cornerstone laid in February 2008, it was declared to be the world's first carbon-neutral, zero-waste and car-free city. The City was originally envisaged to become fully powered with renewables (mainly solar), with all waste recycled, and cars replaced by public transport and rapid personal transit vehicles, placed under a 7-metre platform on which the city would stand. Other sustainability features include reduced use of water, recycled grey water, reduced installed power capacity, and urban planning adapted to the local culture and climate.[79] Masdar City's budget was announced in February 2008 at US$22bn, US$4bn of which would be funded by Masdar and the rest through foreign investment and various financial instruments. The city was expected to add over 2% to Abu Dhabi's GDP and create over 70,000 jobs.[80] Remarkably, Masdar City incorporates a conscious effort at systematically scaling up and integrating advanced renewable technologies.[81] At the time of writing, the city hosted Masdar's headquarters (in temporary buildings) and Masdar's own research institute. The first major foreign business tenant, Siemens, broke ground for its new, LEED platinum/Estidama 3 pearl certified regional headquarters at Masdar City in late 2011.[82] Originally, the city was also designed to attract wealthy expatriates interested in this unique niche of the high-end real estate market.

The Masdar Institute of Science and Technology (MIST), established with the Massachusetts Institute of Technology, was set up in order to attract foreign companies and know-how to the emirate. Research concentrates on a wide range of alternative-energy technologies and policies, and the institute has engaged in a number of partnerships with government, academia and businesses, including research on carbon sequestration with ADCO and on smart grids with Siemens. The first

Master's degree programmes began in 2009, and the institute aims to grow into a 200-faculty, 1,000-student institution by 2017.[83] Another research-related endeavour is an innovation centre of General Electric, as a part of a US$8bn joint agreement with Mubadala, focusing on energy efficiency and awareness-raising.[84]

The utilities and asset management unit (later split into Masdar Power and Masdar Capital) is building up a global portfolio of renewable and alternative energy technology companies. The first investment vehicle was the US$250m Masdar Clean Tech Fund, launched in November 2006 in cooperation with Credit Suisse, Consensus Business Group and Siemens. A second US$265m fund was set up with Deutsche Bank in 2010. The units have, among others, built solar plants with Sener of Spain (including a state-of-the-art 20 MW CSP plant and two other 50 MW CSP plants in Andalusia), as part of the US$1.2bn Torresol Energy joint venture,[85] and invested US$175m in the Finnish wind-turbine manufacturer WinWinD, the largest foreign industrial investment in Finnish history.[86] In October 2008, after Shell had withdrawn from the massive British wind energy project the London Array, complaining of rising costs and low government incentives, Masdar joined the energy companies E.ON and DONG Energy by buying a 20% share in the project.[87] Also, a separate industries unit concentrated on the development and production of photovoltaic (PV) solar energy technologies and energy. The unit's main investment was Masdar PV, a company established in 2008 with the aim of rising to the global top three of the industry. An initial US$230m investment included the construction of a manufacturing plant in Germany.[88]

Domestic production of concentrated solar power (CSP) is another target area. The flagship project in this sector is the Shams 1, a 100 MW plant in the Western region of Abu Dhabi, which is expected to be completed by 2012.[89] In May 2009, a 10 MW photovoltaic system—at the time the MENA region's largest, worth US$50m—was connected to the emirate's grid.[90] Masdar is also working on a second 100 MW CSP plant, Nour 1, and has been considering major expansion in wind energy too, to meet the emirate's 7% renewable energy target. In 2011, the company announced a tender for a 30MW onshore wind farm on the emirate's Sir Bani Yas nature reserve island, which received a go-ahead from the government in early 2012.[91] In 2008, Masdar declared plans to build the world's first commercial hydrogen power plant together with BP and Rio

Tinto, costing US$2bn. The 420 MW gas-powered plant was also envisaged to capture the carbon dioxide from the process and inject it back into depleting oil fields for enhanced recovery, but seems to have fallen victim to hesitation by the Abu Dhabi government regarding the viability of capturing the carbon.[92]

Masdar's carbon unit focuses on carbon capture and storage (CCS) technologies and projects and Kyoto Clean Development Mechanism (CDM) projects. The unit's quantitative goal has been to cut emissions by 30 million tonnes of CO_2 annually by 2020, decreasing Abu Dhabi's carbon footprint by a third.[93] In the summer of 2012, the tentative CDM portfolio of the unit consisted of seven projects (registered or at validation). The emission reductions and financial returns expected from Masdar's first registered seven-year CDM project, the 10 MW PV solar energy plant, are a modest 15 kt of CO_2e, valued at less than US$300,000/year at late 2009 prices. Masdar has also engaged in energy projects and cooperation agreements abroad, including a wind farm and a fuel switching project in Egypt, an emission reduction agreement with Bahrain's National Oil and Gas Authority, and two massive gas pipe leakage reduction projects in Uzbekistan.[94]

In 2008, Masdar announced plans to develop an emirate-level network for enhanced oil recovery that would become the world's largest single integrated CCS project, including 300 km of pipelines. The first phase of the US$3bn project was expected to capture 5–6.5 million tonnes of CO_2 from a steel plant, a conventional power plant, the hydrogen power plant and a future aluminium smelter by 2013, with each tonne adding 2.5–3 barrels of oil to oilfield production.[95] Initially plans did not proceed quite as planned, as ADNOC opted for a more careful approach. In November 2009, Masdar and ADNOC's onshore oil and gas company, ADCO, began testing CO_2 injection in the Rumaitha field. ADMA-OPCO, the offshore gas company, has also been conducting tests at Umm Shaif and Lower Zakum. ADNOC's operating company Gasco and Masdar have also worked on a US$420m carbon recovery and injection project at the Habshan field, but the project partner ADCO's demands for currently unavailable technology, among other things, have delayed the bidding process.[96] Announcements in early 2012 by ADNOC and Masdar on progress in negotiations, however, gave renewed hope of the plans going forward. Masdar also announced plans to tender for capturing up to 800 kt/year from the Emirates Steel plant.[97]

Finally, three other Masdar-related concepts, running since 2008–09, are the highly successful World Future Energy Summit, the European Future Energy Forum, and the Zayed Future Energy Prize. The newest addition will be the Abu Dhabi Sustainability Week, starting from 2013.[98] These are geared mainly at drawing global investor and media attention to Abu Dhabi's new industries and the leadership's major plans and accomplishments.

'For better for worse', 2009 onwards

Starting from 2009 and materializing in 2010, Masdar found itself amidst a major crisis as a consequence of, among other things, the global financial crisis, over-optimistic assumptions, hasty marketing, colossal promises, rushed implementation and, most likely, bad recruitment choices. As Caton and Ardalan have noted in their analysis of Masdar City's predicament, 'vision and innovation driven by scientific goals [ran] up against the pragmatics of the market place'. Mubadala expected quicker returns on its investment.[99] Details of the scale of problems that had been mounting were never revealed to the press despite rumours being rife, and the internally brewing crisis was managed in Dubai-esque style. The financial crisis took its toll, as real estate prices plummeted, making Masdar City's high-tech developments economically unfeasible. Also, the fall in oil prices reduced the amount of government funding available for risky low-return test bed projects. In addition, foreign companies' willingness to commit did not turn out as strong as expected, in part, presumably, because of the global economic downturn.[100]

One of Masdar's major mistakes had been over-reliance on technologies that were not yet ready or could not be implemented on the city site.[101] Technical problems began emerging as early as 2008, as tests revealed that because of the high temperatures and dust, PV solar panels were operating at less than 40% of advertised maximum capacity.[102] According to press sources, the 10 MW plant in Masdar City built in 2009 was actually producing only 2–3 MW.[103] At the solar plant, a group of workers is in charge of manually cleaning dust from the panels, although it is planned that machines will eventually take over this role. Also, delays were caused by the fact that the hired construction force was not appropriately skilled or qualified to use the special materials and technologies with which the Masdar Institute building was equipped. Policy obstacles were also a major hindrance: wind turbines, originally in the master plan, could not be set up on the city area because of restrictions due to the vicinity of the airport.[104] A major impediment to wider-scale implementation of solar energy in the country was the absence of feed-in-tariffs (subsidies) or other types of policies to support solar against fossil fuels. Masdar PV also cancelled plans to construct a solar panel factory in Taweelah (Abu Dhabi) because of the lack of government regulation to support a domestic market.[105] Also, the financial crisis and a sharp drop in the price of a competing technology reduced the competitiveness of Masdar's choice, thin film PV.[106]

2010 was a year of major turmoil for the entire company. In early 2010, in addition to major personnel cuts in Abu Dhabi,[107] two Masdar senior executives, Tassabehji and Awad, left the company, while CEO al-Jaber ordered a six-week rethink of the project's original plans.[108] In May, Masdar PV dismissed both its CEO and COO; in June, Masdar Institute's head of research quit his job, and in July, the Institute's provost resigned for 'personal reasons'.[109] Rhetoric, according to which all visionary projects required constant reviewing, entered the picture, as rumours regarding Masdar's future grew. According to al-Jaber, 'Masdar is by definition a work in progress. Our development activities constantly inform the way we manage and evolve our projects'.[110] Over the summer, he assured the press and international audiences that the company was not scaling back its plans and the UAE was on track to deliver the 7% renewables target.[111]

The year 2010 was characterized by reviews of the entire project, most importantly of the Foster + Partners-designed master plan. In January 2010, Masdar officially dropped its aim to complete the City development by 2016.[112] A phrase often repeated by Masdar executives in 2010 became the promise 'the vision remains the same'.[113] In the summer of 2010, a high-level executive felt that Masdar PV had 'lost a lot of trust and drive' because of CEO al-Jaber's need to 'move very carefully'.[114] The new master plan of Masdar City was announced in October 2010, with revised completion dates: from 2013 to 2015 for the first phase, 2020–2025 for the entire project.[115] In late 2010, rumours circulated that Masdar was possibly considering switching the aim of making the City 'carbon neutral'—a substitute for the even more ambitious 'zero-carbon' of the very first plan—to 'low carbon'. This, if it happened, was regarded by UAE-based renewable energy experts as disastrous for Masdar's international image.[116] Indeed, indications of this were given by the Foreign Minister, Sheikh Abdullah, in his speech to the Cancún climate conference in December 2010, in which he described Masdar City as 'a cutting edge low-carbon urban centre'.[117] The same month, the budget of the City was cut by over a fourth, from US$22bn to US$16bn. Cancellations included, among other things, a pod transport system, elevating the entire project on a platform, and a plan to install solar panels on all roofs. Also, it was decided that construction would proceed in a more phased manner, by neighbourhoods.[118] Important cuts continued in 2011: in September, a re-evaluation was announced for the company's headquarters, originally promised as part of the UAE's IRENA bid to be energy positive. Now, the project would 'be built in phases, in line with market demand and economic conditions at the time'.[119] Finally, in November 2011, Masdar announced another major round of personnel cuts, of 9%, as part of a broader trimming in government-owned companies in Abu Dhabi.[120]

Nevertheless, there were positive developments too, as many projects still proceeded as planned. In the autumn of 2010, the first part of the Masdar Institute in Masdar City opened for students. In June 2010 Masdar announced a US$600m deal with France's Total and Spain's Abengoa to build the Shams 1 (100 MW solar plant) by 2012.[121] And as a sign of endorsement by the Emirati community, the first private donation, 1 million dirhams (US$270,000), was given to the Masdar Institute by a national in late 2010, only to be topped by another, US$1m donation in 2012.[122] Moreover, the perception-changing impact of Masdar, together with the increasing attractiveness of the solar industry, made two other Abu Dhabi government-owned investment companies interested in the sector: the International Petroleum Investment Company partnered with German and US companies to build CSP stations in the UAE, South Africa and Spain; and the Abu Dhabi National Energy Company (Taqa) formed strategic partnerships with French and Spanish companies to develop wind farms.[123] In August 2011 Masdar, the US Department of Energy and the research institute RTI entered into a US$3m collaboration agreement on CCS research, with US$700,000 coming from Masdar.[124] In a medley of foreign aid, personal interests and image repair, Masdar also announced the construction of an 18 MW onshore wind farm on Mahé island of the Seychelles. Not coincidentally, President Sheikh Khalifa bin Zayed's new palace on the same island had been the object of compensation demands of millions of US dollars for the local environmental damage it was inflicting.[125]

Masdar's insiders sought to portray the re-evaluation in a positive light, arguing that this was a sustainable move, as it enabled the company to separate successes from areas that needed development.[126] Most important, however, as a sign of recognition of Masdar's central role in the country's emerging climate policy (and thus providing an official channel of influence for Masdar in it), CEO al-Jaber was, in February 2010, named the head of the new Directorate of Energy and Climate Change, established in the Foreign Ministry, with the titles Assistant Minister for Foreign Affairs and Special Envoy for Energy and Climate Change. International recognition followed in 2012 when al-Jaber was awarded the UNEP's prestigious Champion of the Earth award—the same granted to Sheikh Zayed seven years earlier.[127]

Evidently, by 2011, the green utopia and its long and many tentacles had extended too far into Abu Dhabi's political and economic fabric for Crown Prince Mohammed to fully withdraw support, despite the endless-seeming rollercoaster of successes and failures. Visiting the City, with a massive entourage, he commended the company for being 'on the right track and achieving steady progress'.[128] At least the forces behind the Masdar Initiative certainly learned the hard way the difficulties of delivering fast results and cash, as typically expected in a rentier economy in this highly knowledge-intensive and technology-dependent sector; and recognized that oil revenues are still king, in determining the time and pace of diversification efforts. In private discussions, Sheikh Mohammed's support, according to assurances by the Energy Minister al-Hamli, remains very strong.[129] Abu Dhabi's government, however, is known to include elements that do not consider Masdar as important. These supporters of the old economy, according to some views, will need to see before believing that the initiative can play a role in economic diversification.[130]

Masdar's functions

Because it is so recent, Masdar's future role and its functions in Sheikh Mohammed's vision of Abu Dhabi still remain undecided. The company's official narratives, repeated in commercial material and by Masdar's representatives, can be grouped under three themes: its myth of origin, its economic function and its environmental motivations. First of all, according to the company, Masdar reflects Sheikh Zayed's vision and is a natural continuation of his environmental legacy; also, Abu Dhabi's experience in the oil industry gives it a comparative advantage in alternative energies. Secondly, Masdar is described as an important part of Abu Dhabi's diversification strategy: the industries it incorporates will establish a new economic sector and transform the emirate into a global leader in and exporter of sustainable energy technologies, reversing the flow of high technology and ultimately transforming Abu Dhabi into a post-oil economy by the 22nd century. Thirdly, the initiative's spokesmen and government officials have repeatedly stated that Masdar shows Abu Dhabi's serious commitment and regional leadership in both climate change mitigation and energy security. They also stress Masdar's complementarity to Abu Dhabi's energy and economic security interests, framing the initiative as proof of a rentier state's ability to engage in long-term planning and an oil exporter's ability to be green.[131]

Clearly, the functions of the Masdar Initiative are as multidimensional as the project itself. The fact that the impulse to establish Masdar came from inside Mubadala indicates that it is, first and foremost, another part of Sheikh Mohammed's new economy, aimed at diversifying Abu Dhabi's fossil-fuel-based rentier state. However, some of Masdar's elements, such as the massive CCS projects, are likely to provide important support to the old, oil-based rentier economy, while renewable energy and real estate will boost Abu Dhabi's 'neo-rentier' economy.

In order to create a sustainable knowledge-based sector of the economy, hiring locals and building up permanent human capital are crucial. In 2011, the share of Emiratis among the Masdar Institute's students had climbed to 43% but, despite a promising start in staff Emiratization, reflecting the only emerging domestic know-how in the area, a large part of Masdar's key staff still consisted of expatriates.[132] It will also remain to be seen how eco-conscious Emiratis will be by the time the first residential units at Masdar City are finished. Awareness has risen rapidly, but according to a survey conducted by Masdar, while some Emiratis were very interested, others saw that it was not meant for the nationals' needs and complained that Masdar is more outward-focused than inward.[133] This serves as an indication of Masdar's strong regional prestige orientation and its catering to Western audiences' value systems, but it also shows that Abu Dhabi's 'monarchy' had made a 'pre-emptive strike' by taking the leadership in environmental sustainability before the younger and more environmentally savvy generations grow older.

Masdar is the region's largest single cluster of investments in renewables and other alternatives to fossil fuels. But as the fall of oil prices in the late 2010s revealed, the initiative continues to be heavily dependent on the medium-term success of Abu Dhabi's rentier economy, as well as political support from the upper echelons of power. Moreover, without Abu Dhabi's massive windfall profits and surpluses in the 2000s oil boom years, the funds for Masdar would have never been envisioned in the first place. On the positive side, the decades' experience from the fossil-fuel industry and scaling up of technologies might provide important lessons and support for the company.

Since its establishment, Masdar has not only played the role of a company, but has developed into an influential institution, capable of influencing policy decisions in a number of sectors and policy areas, including urban planning, construction, industries, domestic energy policy, external climate policy, and, not least, promotion of environmental sustainability—partly by setting an example—on Abu Dhabi's entire policy agenda.

Future prospects

If trust and confidence at the highest levels are maintained through successful performance, Masdar has the potential to reinvent Abu Dhabi as a solar energy power, in terms of both technology and energy exports. However, in the medium term, Masdar's role in domestic energy security will remain marginal. While the share of renewables in Abu Dhabi's domestic electricity capacity is set to rise to 7% by 2020 (or roughly 1.5 GW), under the existing subsidy regime solar energy is considered mainly as viable for summertime peak-load supply only.[134] However, given that the emirate's first nuclear plant is expected to be operational at the earliest around 2017, Masdar's solar energy installations could still play a role in alleviating the federation's energy crisis, if promptly implemented.

Although the company does not have a defined authority in the UAE's international-level climate policy, climate change is one of the central themes in the 'Masdar narrative'. One of the initiative's original drivers was the government's wish to show awareness of the UAE's high carbon footprint.[135] The company has also played a decisive role on the emirate's domestic climate policy agenda: the 7% renewables target was the most ambitious and clearest tangible formulation of a domestic climate policy in the Gulf, and still is. Also, in a decade or two, CCS projects pushed forward by Masdar will potentially yield significant domestic emission reductions and set the emirate among the regional leaders in this field of future energy technologies.

Accusations of Masdar being merely a green façade, a Disneyland,[136] of an 'island of efficiency'[137] that will allow the rest of Abu Dhabi and the UAE to continue business as usual, are however pertinent, although decreasingly so. Many renewable energy projects are being pursued beyond the boundaries of Masdar City, a comprehensive sustainable building code has been set at the emirate level, and in a confluence of domestic energy security imperatives and Masdar's lobbying, support for renewable energy subsidies has been growing visibly. Both Abu Dhabi and Dubai have been studying feed-in-tariffs. In late 2011 the UAE's Minister of Economy, Sultan al-Mansouri, gave his support to installation subsidies for renewable energy projects, similar to a 'green payment' from the Abu Dhabi government to Masdar's Shams 1 CSP plant.[138]

Notably, the UAE's energy demand is still expected to grow significantly, and cutting fossil fuel subsidies, instead of adding new ones, is still not openly discussed in Abu Dhabi. Massive industrial and real estate projects are being planned and constructed, which will further increase domestic emissions, especially if fossil fuels will be used to power them. Without working on energy efficiency and curbing demand through subsidy cuts, this demand growth will both undermine any positive impact of renewable and clean 'future energies' and continue to challenge domestic energy security.

The case of nuclear energy[139]

While Masdar was driven mainly from within the emirate, its linkages to external drivers are evident in its key themes and image strategy, firmly anchored in a globalized, Western-influenced discourse on climate change and environmental sustainability. Abu Dhabi's nuclear programme is a distinct case in that the role of foreign actors in pushing the local rulers towards this energy choice has been much more visible and tangible. Foreign governments and companies have also played a key role in the evolution of related policy, regulatory and institutional developments and diplomacy.

Against the expectations of many, starting in 2007 the foundations of the world's fastest emerging civilian nuclear power programme were laid out in Abu Dhabi, with the crucial help of external powers. Finding itself amidst an energy crisis and with no prior domestic technical expertise, the government decided to outsource the entire programme. The fast implementation schedule, which envisages the completion of four 1.4 GW reactors by 2020, will require strong political will, solid financial support and a high level of engagement from the top of the ruling family. Owing to international concerns over proliferation, Abu Dhabi has chosen to emphasize transparency, non-proliferation, and trust-building as the core of its internationally communicated intentions. The support of key international actors, principally the United States, has enabled Abu Dhabi to access the best technologies and start implementing the programme at record speed. Moreover, the US strategy to portray the country's *modus operandi* as a 'model' for the region, and the world, has provided the government with important regional prestige opportunities. Alongside a detailed analysis of the first years of the programme, the case study—because of the strategic nature of nuclear power and the important linkages to regional security and foreign policy, as noted—also shows the direct influence that foreign governments retain in some areas of Abu Dhabi's and the UAE's policy agendas.

The nuclear choice

According to a high-level official, the government of Abu Dhabi first started considering nuclear energy in 2006, when a local inter-agency energy planning committee was mandated to produce a more accurate estimate of future demand for electricity and water in the emirate. Soon afterwards the mandate of this Energy Working Group was extended to the entire federation and broadened to chart the feasibility of different technological options to meet future demand.[140] Simultaneously, from around 2007 France and the United States became increasingly involved in promoting nuclear energy to the UAE. President Sheikh Khalifa's visit to France in July 2007 laid the foundation for bilateral nuclear cooperation. On the French President Sarkozy's initiative, the two agreed to finally put into effect a 1980 agreement on the peaceful use of nuclear technology. An American nuclear consultancy also began working for a branch of the Abu Dhabi government from late 2007. In March 2008, after a rapid series of consulting, negotiations and bilateral agreements, the UAE's Supreme Council of Rulers approved a memorandum, submitted by the Foreign Minister Sheikh Abdullah, which explained the motivations and principles of the country's potential plans to develop a peaceful nuclear energy programme.[141]

The contents of the memorandum were released in April 2008 in a white paper titled *The Policy of the UAE on the Evaluation and Potential Development of Peaceful Nuclear Energy*, which stressed the exclusively peaceful nature of the UAE's intentions, alongside the need to develop additional sources of electricity to meet increasing demand. It also underscored the principle of maximum transparency and renounced domestic enrichment of nuclear fuel on the basis of economic unfeasibility and international proliferation concerns. The rapidly rising demand for electricity, which according to a study commissioned by Sheikh Khalifa would increase by 165% by 2020 (see chapter 4), was presented as the main factor behind the decision. The study showed that natural gas would be able to supply only half the needed capacity and renewables 7%, at most. Moreover, the Nuclear White Paper noted that despite their 'logistical viability', burning crude oil or diesel for electricity production would 'entail extremely high economic costs, as well as a significant degradation in the environmental performance of the UAE's electricity sector'. Similarly, coal was discarded as unviable, according to the paper, because it was environmentally unfriendly and involved 'thorny issues related to security of supply'.[142] An additional concern was dependence on imported coal, which would amount to 'multiple shipments every week' through the geostrategically critical Strait of Hormuz,[143] which in theory at least could be blocked by Iran. In addition, the policy paper noted that 'stacked against the above options, nuclear power-generation emerged as a proven, environmentally promising and commercially competitive option which could make a significant base-load contribution to the UAE's economy and future energy security'.[144]

Establishment of the programme

The seriousness of Abu Dhabi's plans was confirmed in June 2008, when a call for initial bids for the construction of the first four reactors, through joint ventures, was announced.[145] In July 2009, the local implementing authority presented a short list of three consortia (French, US-Japanese and South Korean),[146] and in December 2009 KEPCO, the South Korean, was chosen with a bid of US$20.4bn.[147] Interestingly, while there were important trade links between the two countries, South Korea was obviously not chosen for geopolitical motivations but rather, according to leaked US embassy cables, for the lowest price and expected reliability of delivering on schedule.[148] The four reactors the Korean consortium has agreed to build in Braka, over 300 km from the city of Abu Dhabi,[149] have an extremely ambitious schedule, given that in 2008 the federation still not did have any of the needed legal, institutional or infrastructure framework in place. Preparatory groundwork in Braka began in early 2011, with 1,000 Koreans already working on the site.[150] Construction of the first plant started in July 2012, and work on the second is expected to begin in 2013.[151] The plan is to have the reactors operating in 2017–20, one each year. According to a UAE nuclear regulator's estimate, 'everything needs to work for the schedule to hold'.[152]

According to the Emirates Nuclear Energy Corporation (ENEC), the factors supporting the nuclear choice were economics, security of supply, environment, and industrial development.[153] While the possibility of military motives influencing the UAE's choice to start up a nuclear energy programme can never be excluded, the economic justification deserves attention for many reasons. First, several economic motives emerged in the mid-2000s, including diminishing unused crude oil production capacity, growth of the petrochemical industry, the competitive price of nuclear power for base load electricity supply, and the rising opportunity cost of burning oil at home.[154] Secondly, changing assumptions regarding long-term oil prices, which took place around 2005–06, are said to have led to a change of heart in the Gulf monarchies towards nuclear power, as rising energy demand from emerging economies like China and India started to weigh in, pushing international energy demand projections up considerably.[155]

Maintaining absolute transparency and the highest possible security and quality standards is seen as the vital condition for any Gulf state wishing to build a nuclear-energy programme. Accordingly, the five main principles of the UAE's 2008 policy were outlined as: transparency, non-proliferation, safety and security, conformity and working with the IAEA, and working with friendly nations and expert organizations.[156] A number of key elements, such as competent staff and assistance for setting up a nuclear law and the necessary regulatory institutions, were available only from abroad.[157] The UAE's peaceful intentions were repeatedly stressed by the government, along with calls for Israel to sign the Nuclear Non-Proliferation Treaty (NPT)—arguably to please the Arab street—and for Iran to continue cooperation with the IAEA.[158] In October 2009, going beyond the NPT, the UAE issued an energy law that outlawed domestic enrichment of uranium.[159] The country is a member of the NPT (1995) and the IAEA Safeguards Agreement (2003), and in 2009 it signed the Additional Protocol to the Safeguards Agreement, among other key nuclear conventions.[160] On the fuel supply side, an announcement on the winner of a lucrative 15-year deal is expected in 2012.[161] The UAE has supported the establishment of an international nuclear fuel bank as a safeguard for countries that do not enrich uranium domestically, and it has donated US$10m to the American Nuclear Threat Initiative administered by the IAEA.[162] The White Paper also implies that long-term storage of nuclear waste on national soil is not the *preferred* option.[163]

As an indication of Abu Dhabi's full ownership of the programme, apart from the nuclear bid itself, the local nuclear company is Abu Dhabi government-owned and all consulting contracts have been signed by the local rather than the central government.[164] Hamad Ali al-Kaabi, appointed as representative to the IAEA in 2008, has been a key Emirati figure, and the public face for the programme.[165] Based in the Foreign Ministry, he was responsible for the feasibility study and consequent 2008 White Paper, and has been involved in all bilateral discussions with potential supplier countries, acting as the main interlocutor on behalf of the UAE government.[166] The key patron figure of the nuclear issue has been the Foreign Minister Sheikh Abdullah, the public face of lobbying for international approval of the UAE's nuclear plans. David Scott, former regional director of the US National Security Council, and currently the director of economic affairs at the Executive Affairs Authority, is said to be the key expatriate behind the programme.[167] The Abu Dhabi-owned Emirates Nuclear Energy Corporation (ENEC) was set up in 2008 as the responsible implementing authority with an initial budget of US$100m.[168] In October 2008, ENEC appointed the American company CH2M Hill (which also manages Masdar City) as the managing agent of the nuclear programme with a 10-year contract.[169] An Abu Dhabi-based Federal Authority for Nuclear Regulation (FANR) and an international advisory board were also set up, with a senior US regulator leading the former and Hans Blix the latter.[170] Although the nuclear programme will rely heavily on contractor services for technological expertise for the foreseeable future, the government has begun to work on developing domestic expertise through the Gulf Nuclear Energy Infrastructure Institute at Khalifa University, with the assistance of Texas A&M University and US government-owned Sandia National Laboratories.[171]

The external supporters

As an indicator of earned credibility, in 2008 the UAE was already declared by a leading international security institute as most likely to be the first Arab country to produce nuclear energy.[172] The major global suppliers of nuclear technology and fuel have engaged in supporting and promoting the nuclear option to the UAE government, and at the same time sought to win a stake in the multibillion-dollar project. The United States has eagerly used the case as a carrot for Iran, demonstrating the benefits of complying with internationally agreed standards on nuclear development. By taking advantage of this opportunity, the UAE managed not only to gain the confidence of relevant international powers, it also succeeded in pitching them against each other in competing for significant business opportunities.

Negotiations with the French government and the Bush administration in the latter half of 2007 confirmed that both countries were interested in promoting nuclear energy in the UAE. An agreement on peaceful nuclear cooperation with France was signed in January 2008, when three French companies (Areva, Suez and Total) also signed a partnership agreement with Emirati counterparts proposing the construction of two reactors.[173] Understandably, observers have seen the nuclear issue as being linked to the French military base which was inaugurated in Abu Dhabi in May 2009, and is thought to be linked to France's aspiration to secure its commercial interests in the region and even to contain the Iranian threat.[174] The United States, under Presidents Bush and Obama, has been a strong supporter of the UAE nuclear programme. A bilateral memorandum of understanding (MoU) was signed in April 2008, but owing to the transition of power in the US administration, the signing of the so-called 123 agreement was delayed until January 2009.[175] The agreement was finally accepted by the US Congress in October 2009, after delays caused by concerns about non-proliferation issues and rule-of-law and human-rights violations, including a torture case involving

Sheikh Issa bin Zayed Al Nahyan, half-brother of Sheikhs Khalifa and Mohammed.[176] The 30-year treaty allows companies to practice nuclear trade in the two countries, but the United States reserves the right to withdraw if its terms are violated.[177]

The UAE has also signed MoUs with the United Kingdom (May 2008) and Japan (January 2009), and an agreement with South Korea (June 2009). It is reported that as early as 2005, South Korea and the UAE explored the possibility of constructing a small pressurized water reactor.[178] While the major global suppliers have similarly sought to court other states in the region, with Saudi Arabia and Jordan expected to launch calls for construction bids some time in the early 2010s, Abu Dhabi was, in 2012, far from any other Arab state in terms of implementation.

The Emirati-Korean deal has additional benefits for both sides: despite lacking the ability to provide the UAE with security backing like the US or France, the South Koreans were lauded by ENEC's CEO for having 'dedicated a highly experienced team… [and] shown a serious commitment to transferring the knowledge gained from Korea's 30 years of successful nuclear industry operation'.[179] Developments in the area of bilateral cooperation after the Korean consortium won the bid in December 2009 have confirmed the expectation that the cooperation would increase trade deals in other sectors as well: at the end of 2009, Mubadala and the Korean Ministry of Knowledge Economy announced an agreement on promoting joint projects between Masdar and Korean renewable energy and technology companies.[180] In 2010, Abu Dhabi awarded a Korean-Japanese consortium a gas power plant deal worth US$1.4bn.[181] In 2011, the Global Green Growth Institute, an initiative of the Korean government, opened a regional office in Masdar City (which influenced the UAE's green growth plan announced the following year).[182] Moreover, the nuclear project is estimated to require cooperation for close to 100 years,[183] indicating that this was only the prelude to enhanced bilateral relations between the two countries.

The eagerness of suppliers to provide the UAE with the needed technology was undoubtedly directly linked to the strong non-proliferation safeguards the UAE attached to its policy. Everything was done by the book, and beyond. The process through which the nuclear policy was developed included multi-stakeholder consultations at both the domestic and international levels: after determining the viability of nuclear energy for the UAE, the government developed a set of guiding principles, which were later embedded in the White Paper, and conducted a study of international best practices in the industry, as mentioned above. Before formally endorsing the policy, the government engaged in consultations with a number of supplier countries (France, the US, the UK, Russia, China, Japan, Germany and South Korea) and the IAEA. In developing a high-level strategy for pursuing the development of the nuclear programme, the UAE followed the IAEA's guidance and the agency's *Milestones* document.[184] A resulting 'Roadmap document', shared with the IAEA, included a feasible schedule for construction, recommendations on the needed order of actions (with site selection highlighted), and the establishment of the required institutions and passing of the needed legislation.[185] What followed was an impressively well-thought out foundation for a federal framework, in which Abu Dhabi could build its four nuclear reactors, and possibly many more, at a world record speed.

Although the programme had won the confidence of key Western powers, many observers still argue that it could increase the possibilities of a regional nuclear race. Other concerns have included terrorist attacks and domestic political instability. The UAE does not have a completely clean record in non-proliferation; according to US and UN officials, Dubai was a central transfer point for the Pakistani nuclear scientist A.Q. Khan in the illicit sale of nuclear technology to Libya and North Korea in the 1990s. Dubai has its thriving trade with internationally embargoed Iran and hosts many Iranian banks, and in addition is alleged to have been the transit point for supplies of military and dual-use material to the country.[186] As a show of intent, the UAE reported in 2009 that it had closed dozens of Iranian companies and blocked illegal shipments of goods destined for Iran. Also, Abu Dhabi, with its less warm relations with Iran, pledged to put pressure on authorities in Dubai to reassure its external supporters.[187] A stronger export control law in the UAE from 2007 was boosted by amendments in 2008 and the formation of a committee on commodities subject to import and export control in 2009.[188]

Regional geopolitics and prestige-seeking

There is a complex strategic calculus behind the UAE's nuclear programme: security apprehensions concerning Iran's intentions, the three-island dispute, and the Iranian population in Dubai, in addition to the common desire of the US and Abu Dhabi to present a counter-example to Iran's handling of its own nuclear programme. The UAE's head start on Saudi Arabia can be seen as one means of boosting its position within the GCC. While declaring support for the GCC joint viability study, the UAE government's White Paper implicitly affirms both a strong determination to pursue a national programme, independent of the often difficult regional cooperation, and a will to raise its regional status as the first Middle Eastern state to operate a civilian nuclear-energy programme with the full approval of the IAEA.[189] Arguably, Saudi Arabia will be keeping a close eye on Abu Dhabi, as it might perceive the programme as an attempt to weaken its relative power vis-à-vis both the GCC and its Western allies. Also, if external suppliers and their developing country partners replicate the 'UAE model', the country will always constitute a reference point, a 'gold standard',[190] with which other countries' programmes will inevitably be compared.

Future prospects

By the end of 2011, many factors supported a positive outlook on a timely implementation of the nuclear programme: long-term oil price projections, Abu Dhabi's financial surpluses, domestic energy demand pressures, the government's political determination, and the authoritarian political system too. Even Japan's Fukushima disaster did not seem to slow down plans: FANR announced possible changes to design and location, but not to the timeline. Also, staying true to promises on safety, ENEC set up a task force to reflect on 'any lessons learned' from the accident.[191] Since domestic debate on the topic has been non-existent, if not actively discouraged,[192] the government has been able to 'educate' the public and significant domestic opposition to nuclear energy is unlikely to evolve in the future. After the Fukushima events, ENEC initiated a series of community forums to engage with potentially concerned citizens.[193] The company also conducted an acceptability survey of 750 residents, and reported that 85% of the respondents viewed 'peaceful nuclear energy [as] important for the nation'.[194] International support for the programme, in turn, will be secured as long as the UAE keeps to its principles of complete operational transparency, safety, security and cooperation with the relevant international non-proliferation bodies. Although opinions on nuclear energy are generally divisive, there are clear climate change benefits: while international pressure to cut greenhouse gas emissions was not the main motive for launching the ambitious programme, if implemented the nuclear energy capacity will push the country significantly towards lower carbon intensity and lower per capita emissions.

The main domestic challenges for a timely and successful implementation arguably arise from the rentier mentality of the state and its citizens: challenges regarding human resources, institutional infrastructure and wide-scale implementation, which all require consistent and long-term planning and implementation. A major sustainability issue will be the Emiratization of the fully outsourced programme. An FANR official has estimated that during the operation phase, around 500 people will be needed to run the programme. In addition to perhaps 10,000 construction workers, a couple of thousand people will be needed in the establishing phase.[195] In addition to this, the regulatory authority, with close to 100 employees in 2009, might also have to wait a long time before being able to hire a sufficient number of skilled nationals. Also, as with Masdar, while acquiring technical expertise only requires international approval and interest, the success of the UAE's nuclear-energy programme will require the sustained political support and financing of key ruling-elite members so as to deliver on the grandiose promises laid out in the early master plans. As of 2010, the nuclear energy programme seemed to have escaped Masdar's 'fate' of financial crisis-related delays, indicating its immediate importance for the government.

Along with the French military base, the UAE's withdrawal from the GCC currency union in 2009,[196] and the various dimensions of the strategic branding of Abu Dhabi by Sheikhs Khalifa, Mohammed and Abdullah bin Zayed, the nuclear programme should also be understood as a part of a so far extremely successful external strategy to raise Abu Dhabi's profile and gain prestige both among regional peers and internationally. Abu Dhabi's prestige-seeking, however, might still be set back by a number of issues including the authoritarianism of the political system and the serious problems with freedom of speech and rule of law (including the torture case mentioned above and the more recent arrests of tens of UAE citizens); Dubai's past as A.Q. Khan's transit point and its close relations with Iran; tensions with neighbouring countries (which might in the future prompt the UAE to leave the NPT); the preferential treatment loophole (a minute in the 123 agreement, which grants the UAE the right to demand renegotiation of the agreement should better conditions be granted to any other state in the Middle East in the future); and the uncertainty over the adoption of the UAE's model: Jordan and Saudi Arabia, for example, are not looking to give up their right to domestic enrichment.[197]

6

QATAR'S NATURAL SUSTAINABILITY COMPLEX

Qatar, the owner of the world's third largest natural gas reserves, has since the mid-1990s gained world renown for its innovative branding endeavours and unique foreign policy alignment. On the rentier side of the economy, early investments in liquefied natural gas production capacity have brought unprecedented wealth and economic stability. The plentiful natural gas has also been sufficient for domestic demand, making Qatar the only Gulf monarchy that did not face gas shortages in the late 2000s and early 2010s. The unprecedented development of the 2000s did not bring a sense of scarcity, nor did the global economic crisis significantly slow down economic growth. As demonstrated in the following chapters, the picture emerging is that of a runaway train. First, Qatar's natural unsustainability is not producing the kind of immediate resource scarcities that is pushing its neighbouring states towards alternative energies and energy efficiency and conservation. Secondly, the current leadership's external role perceptions and internal legitimacy strategies are, unlike in the UAE, not aligned in a way conducive to pursuing the climate or environmental sustainability agendas in a profound, transformative way. Qatar's challenge in the 2010s will therefore be managing not scarcity, but abundance. Despite a considerable research emphasis, goodwill rhetoric and a number of mega-size prestige projects, the sufficiency of natural gas for domestic consumption combined with the small national population and record-high external rent revenues will significantly slow down progress in the area of natural sustainability, portending a bleak outlook for future environmental sustainability.

As the 2000s drew to a close, power in Qatar remained concentrated in the hands of Emir Hamad bin Khalifa Al Thani and a handful of his closest allies and family members, none of whom exhibited a strong personal interest in climate change mitigation or environmental sustainability in itself. Also, environmental institutions in Qatar were weak and lacked capacity to take on such a multifaceted challenge as climate change. The beginning of a gradual revolution in the international energy economy was, however, not lost on the elite. Nor were environmental 'unsustainabilities', created by the rentier bargain, left unexposed. What emerged from Qatar, endowed with time and resources, at the turn of the 2010s was a largely uncoordinated, piecemeal approach to all these pressures. Only a few more articulate, promising responses began taking form, largely through the technology development agenda of Qatar Foundation and the overarching food security vision of the Qatar National Food Security Programme.

As in the case of Abu Dhabi, the following analysis of Qatar's natural unsustainabilities and responses to climate change is spread over two chapters, the first of which examines the structural drivers of natural unsustainability in Qatar's political economy and its environmental and climate change governance. Then, after an examination of the physical and economic challenges of climate change to Qatar, chapter 7 analyses the role of key elite members in driving some of the late 2000s' most important domestic reactions and responses to the issue of climate change. The main case study examines the research and development initiatives under Qatar Foundation, including its Qatar Science and Technology Park, which represents the emirate's most advanced cluster of alternative energy and environmental sustainability projects. The chapter also points towards Qatar's plans for sustainable agriculture and food security, which could potentially bear broader implications for nationwide implementation of alternative energies and more sustainable resource use. While the chapter illustrates prime examples of neotraditional survival balancing of a rentier ruling elite, both limited and aided by the rentier state structures and the local decision-making structures and dynamics, it also serves to demonstrate how international trends, such as clean and green technologies, and external image strategies and branding have influenced the course of development choices and policies in the case of Qatar.

Fossil and non-fossil economies

Qatar is the ultimate strong rentier state, as defined in this volume. It is an extremely wealthy, small high-income developing country, with the world's third largest proven natural gas reserves and the 13th largest oil reserves (13.5% and 1.9% of world totals respectively). In 2006, after less than a decade of exports, it surpassed Indonesia as the world's largest LNG exporter.[1] In the 2000s, according to Ernst & Young, Qatar's economy grew at an impressive average rate of 13% per year.[2] Growth has been predominantly government-led, while Qatar, arguably because of the LNG sector and the 'relative strength' of gas prices, also managed to weather the global financial crisis exceptionally well, with an estimated growth of 11% in 2009.[3] Between 2010 and 2014, Qatar is planning to invest US$31bn in gas infrastructure development and other energy and industry projects.[4] In 2010, although ranking only 59th in the world in terms of the size of the economy, Qatar's GDP per capita was US$88,200–179,000 (depending on the source), which placed the country as the most affluent in the world.[5] Again, as in Abu Dhabi, the GDP/capita of an average Qatari national is several times higher. In 2011, Qatar National Bank estimated Qatar's oil and gas revenues at US$79,000 per national.[6] In essence, the state's main development challenges are the same as those of the other GCC monarchies, including sustaining the economy through diversification and increasing the skills and competitiveness of the national workforce.

Oil and natural gas have been the prime contributors to Qatar's economic development, although the relative importance of the former is decreasing.[7] In 2002–08, according to International Monetary Fund statistics, the oil and gas sector's share of GDP was on average 58%, while in 2008 hydrocarbon exports totalled 92% and petrochemicals 4% of all exports.[8] According to the Qatar National Bank, in 2010 the country's oil exports totalled US$20.4bn.[9] Natural gas plays an extremely vital role in the country's present and future development. Qatar's LNG export programme was initiated during the 1997 fall in oil prices, and already in 2008, following massive capacity expansions, total gas revenues surpassed oil revenues for the first time.[10] Between 2009 and 2010 alone, exports grew from US$26.0bn to US$39.0bn, according to the Qatar Central Bank.[11] As in the case of other GCC oil monarchies, the comparative advantage achieved from cheap energy sources has supported development of heavy industries, like iron and steel, petrochemicals and cement. The major energy and heavy industries are concentrated in the two industrial cities north and south of Doha.

Qatar's only sovereign wealth fund, the Qatar Investment Authority (QIA), with an estimated size of US$85bn in 2011, was established in 2005. Owing to its young age, it is considerably smaller than similar funds of other GCC OPEC monarchies. The rather opaque QIA invests both internationally and domestically (non-energy assets) with an aim to increase economic diversification in Qatar by the late 2000s.[12] The QIA also owns the Qatari Diar Real Estate Investment Company (Diar), which invests in Europe and increasingly in Asia. Famous investments in Western key institutions include the purchase of 20% of the London Stock Exchange in 2007 and a 7% share of the Volkswagen Group in 2009.[13] In 2008, with the aim of boosting Qatar's food security, the QIA established Hassad Foods, which invests in agricultural, livestock and food companies, and it has since then focused increasingly on commodities, food, energy and water.[14]

Diversification from oil to natural gas had already begun in the early 1990s, and in early 2010 MEED reported US$230bn worth of projects either planned or under way.[15] Although the expansion of revenues from the LNG sector will slow down from 2011 onwards owing to a moratorium on Qatar's largest field, expected to last at least until 2015, the government expects growth in services, manufacturing and construction to maintain a nearly double-digit GDP growth. Also, in December 2010 Qatar won the 2022 FIFA World Cup bid, which will create further infrastructure needs and speed up the implementation of existing ones. The government, nevertheless, recognizes that dependence on hydrocarbon revenues 'is set to continue' and that intra-GCC competition in the few attractive non-hydrocarbon niches is fierce.[16] Many areas of Qatar's diversification strategy interlink with its foreign policy strategy. Branding, with its economic diversification aspects, is particularly apparent in the areas of media, air industry, conferences and events, sports, construction and real estate, and education and healthcare (tourism).[17]

Qatar has also successfully increased private and foreign investment in its non-energy sectors: in 2009, the United Nations Conference on Trade and Development ranked it as the world's thirteenth largest foreign direct investment recipient.[18] In addition to economic diversification, government spending is heavily focused on the needs of the nationals; it drives domestic development and allocates wealth among the small national population. In the fiscal year 2008–09, 32% of Qatar's budget was earmarked for infrastructural projects, but 21% went to education and 10% to healthcare.[19]

Since the late 1990s, the government has also initiated a number of heavy investments in gas and other infrastructure and real estate mega-projects.[20] Major projects include the US$26bn Qatar Railway (expected by 2020); the Doha International Airport (US$2.8bn, 2013); the New Doha Port (US$7bn, 2027); and a power plant (US$5.0bn, 2016).[21] In 2004, like many of its neighbouring monarchies, Qatar passed a law (No. 17/2004) that allowed the right to usufruct for non-nationals for a period of ninety-nine years in a number of pre-determined areas. Two mega-projects in the real estate sector are taking advantage of this possibility; the 4km^2 US$14bn Pearl-Qatar and the US$7bn Lusail City development are together expected to provide housing for 240,000 wealthy buyers.[22] Heart of Doha (renamed Msheireb in 2010) is a prime example of a neotraditional project; advertised as reclaiming Qatari identity and tradition and developed by Dohaland, under the Emir's wife Sheikha Mozah, it includes 226 new buildings in the old centre of Doha with a price tag of US$5.5bn.[23] Also, a US$2.6bn Energy City, part of the Lusail development and aimed at hosting international energy companies, was planned in the late 2000s, but seems to have been postponed.[24]

Through the Qatar Foundation, founded in 1995 by Emir Sheikh Hamad bin Khalifa Al Thani and his wife Sheikha Mozah bint Nasser, the rulers hope to transform Qatar into a knowledge economy. The cornerstone of this project is Education City, which hosts branch campuses of prestigious American universities such as Texas A&M, Georgetown and Carnegie Mellon and has a budget of several billion dollars (US$3bn in 2011 *alone*, according to some sources).[25] In addition, the broad, mushrooming mega-initiative embodies strong elements of branding, educational tourism, and the effort to create a new economic sector. Despite the ambitious diversification plans, gas abundant Qatar has not made a similar choice to Abu Dhabi's for heavy investment in alternative energies as a new economic sector. However, a heavy and relatively coordinated emphasis on R&D in a variety of energy and water technologies and solutions is emerging in many research institutions and companies under the Qatar Foundation's umbrella.

Population pressures

From a small town-size total population of 45,000 in 1960, Qatar has in five decades attained a medium-size capital, a number of major industries and a population of 1.7 million. Of this, roughly 14% are estimated to be national, all of them tied in a dense web of mostly material citizenship-related benefits.[26] Since 2005, when the World Bank estimated the country's total population at 885,000, annual growth rates have been over 10%. As in neighbouring monarchies, the rapid population growth since the 2000s has been a direct consequence of Qatar's vast hydrocarbon wealth and the massive diversification drive combined with a minuscule local population, unwilling to work for the private sector and often without the skills to match the market's needs. The number of low-wage workers, coming mainly from South Asia, has been growing particularly fast, both absolutely and proportionally: in 2004, manual labourers accounted for 25% of the population, but in 2008 their share was at 57%. The numerous construction and megaprojects have created a vicious circle of increasing housing demand. In 2010, some analysts predicted that the completion of many mega-projects in the early 2010s would slow down the rate of immigration, but considering Qatar's promises as host of the 2022 World Cup alone, including twelve cooled football stadiums and a high-speed rail and metro system, a major shift looks unlikely.[27] Owing to these projects and the large, volatile share of construction workers, future demographic growth projections are extremely difficult to make. Even Qatar's long-term urban master plan, long in the making, is not based on any future population estimates, unlike that of Abu Dhabi.[28]

There are Qatarization policies, but clearly it was not possible to enforce these in the 2000s,[29] and it is unlikely that foreign labour will be made redundant by them, nor by natural population growth of Qatari nationals, for many decades. While the ratio of males to females among the Qatari adult population is even (1.03), among non-Qataris, because of the large male labour element, it jumps to 5.78 males per female (in 2008).[30] Qatar's national population, however, is growing. UN estimates suggest that the average natural increase rate of the population in 2005–08 was 10.0–11.5 per 1,000, equating to around 10,000–13,000 persons per year; this, arguably, is roughly equal to the number of new jobs the government should be creating annually for Qatari nationals in less than two decades' time. So far, owing to the massive financial resources, it has not been an impossible task to employ the willing and able from among the small national population, estimated at 250,000 in 2011.[31] The rate of unemployment among Qataris is, however, high: only a third of all working-age females and two-thirds of males are employed. In 2008, according to the Statistics Authority, the share of Qataris in the total workforce was 12%. The public sector, one of the main channels of welfare allocation, is the main employer of nationals, representing 58% of the government departments' workforce. In 2008 the private sector, which employed 78% of the total labour force, was composed almost completely (99.5%) of non-nationals. Only 7% of employed Qataris worked for the private sector.[32]

Qatar's political system is a *de facto* absolute monarchy. Like the UAE, it has repeatedly been classified by the Freedom House survey as 'not free', and as authoritarian by the Economist Intelligence Unit survey; the latter has described the state of Qatari politics as a 'near-total domination' by the Al Thani family.[33] Since its independence in 1971, Qatar has held three municipal elections (in 1999, 2003 and 2007). Following the Emir's repeated promises (since 1998), and a new constitution (2005) which implies parliamentary elections, the first Majlis al-Shura elections are expected for 2013.[34] In 2011, Qatar remained the only Arab country yet to see any domestic calls for increased public representation or political reform. Acting pre-emptively, ahead of the curve, in September Emir Hamad raised all Qatari government employees' and pensioners' salaries by 50–120%, in a move that directly clashes with the GCC states' objectives to increase nationals' participation in the private sector.[35]

With the small national population both materially and immaterially co-opted, the top ruling elite seems to continue to face little or no pressure to increase political liberalization. Explanations for this success suggest the same factors and strategies employed by Abu Dhabi, but with differences in depth and nuance: autonomy of the state due to oil/gas revenues; the rentier bargain and related patronage networks; small size and homogeneity of the national population; immaterial legitimacy resources (such as political modernity) created and sustained by the ruling elite, supported by the state coffers; state-building through institutions staffed by Emir loyalists and the establishment of a direct line of succession; and, in general, clever power balancing at all levels, from regional to intra-elite.[36]

The old energy

With regard to its oil sector, Qatar largely confronts the same issues as Abu Dhabi—uncertainty over international demand and price, and domestic production capability and capacity—the main difference being Qatar's smaller reserves. The country's first oil well was drilled in 1939, but exports only began after World War II in 1949. After peaking in the late 1970s, production rates emerged again in 1991 as a result of a new discovery, al-Khaleej.[37] Production reached a record high at 1.57m bbl/d in 2010 (equal to a little over half of Abu Dhabi's production). In 2010, Qatar accounted for 1.7% and 3.6% of global oil and natural gas production, respectively.[38] The existing reserves, however, are maturing and enhanced oil recovery is being considered for various fields.[39] Qatar's remaining oil reserves in 2010 were estimated at 25.9bn bbl[40] and the remaining reserve life at 37–55 years.[41] Similarly to the moderate OPEC producers Saudi Arabia and Abu Dhabi, Qatar has indicated that its 'right price' is in the US$70–80/bbl range, while in 2011 the World Bank placed Qatar's break-even price at only US$40/bbl.[42] The national oil company, Qatar Petroleum (QP), established in 1974 and nationalized in 1976 in the wake of the oil crisis,[43] is fully owned by the Qatari state. In addition to QP's own operations, oil is produced under development and production-sharing agreements with foreign companies.[44] The main market of Qatari oil is Asia, with Japan as the largest importer, as for Abu Dhabi.[45]

Because less value was given to the lower-priced natural gas, its exploitation began only in 1991, mainly prompted by the need to find a long-term replacement for the maturing oil reserves. Qatar's current ruler Sheikh Hamad is generally described as the mastermind of the strategy to develop the country's natural gas resources into a financial and political asset, both to ensure continued flows of external rent after the domestic oil peak and to secure international interest in the country's stability.[46] Since the 1970s, Qatar's gas production had been growing steadily, but in the 2000s it exploded, from 23.7 bcm in 2000 to 116.7 bcm in 2010 according to BP, representing a fivefold increase in a decade. Breaking another record, in 2008 tiny Qatar, owing to exports via the new Dolphin pipeline, accounted for the world's second largest increment in gas supply (36% of global total).[47] Because of the small population and energy demand Qatar can still export most of the gas and oil produced: in 2009, the country's energy self-sufficiency ratio (5.9) was over twice the Middle East average (2.7), according to the IEA.[48]

In 2010, small Qatar was estimated to have some 25 tcm of proven natural gas reserves, which are expected to last over 200 years, at current production rates.[49] This has led many to dub Qatar as the 'Saudi Arabia' of LNG and natural gas exports.[50] Indeed, Qatar's gas reserves are equal to approximately 160bn bbl of oil, or six times its oil reserves.[51] Qatar has planned its gas capacity expansions accordingly. Most of the gas is located in the gigantic North Field, discovered in 1971. However, experts point to 'profound' uncertainties relating to the field's total volumes, potentially affecting Qatar's ability to deliver in the longer term. A moratorium placed in 2005 on the field, for studying its optimal development and also to determine possible damage to long-term productivity inflicted by current exploitation, is currently expected to last beyond its next review in 2014.[52]

Qatar's plentiful gas resources and small population place it in a strong position in terms of long-term economic sustainability. Unlike the situation in the other OPEC GCC states, the country's gas reserves provide a buffer for potential losses incurred from decreases in the global price or demand for oil, both short-term and long-term. Qatar also stands to win in relation to the global shift to lower carbon fuels. Furthermore, it has several options for marketing its gas: LNG, pipeline exports through Dolphin, gas-to-liquids (GTL), and domestic markets.[53] About 70% of Qatar's exported gas is currently shipped as LNG and the rest is piped to the UAE and Oman. As of 2011, the two LNG operators, RasGas and QatarGas, had fourteen trains online, with a total production capacity of 106 bcm.[54]

Qatar's LNG is exported mainly to Asia and Europe: Japan, South Korea, Japan and India (57%); and Belgium, the UK and Spain (33%) receiving the largest shares in 2009.[55] In 2009, Qatar expanded its export portfolio to China and the UK, and by 2010, it was supplying the latter with 15% of its total gas demand.[56] Given that Asia's hydrocarbon consumption is projected to keep rising for the coming decades, the current trade relationships, including those with Europe, place Qatar in a good long-term market position, despite the early 2010s gas glut caused by new discoveries of unconventional gas in North America. Beyond LNG, Qatar also aims at a considerable expansion in its petrochemical industry and has an ambitious gas-to-liquids programme, operated under two joint ventures, Pearl GTL and Oryx GTL, that together produce 170,000 bbl/d of GTL and other products.[57]

Qatar's domestic energy consumption, similarly to the other GCC states, has since the early 1950s been dominated by natural gas and oil. Since the establishment of the local Department of Water and Electricity in 1954,[58] power and water demand have been growing fast. Owing to the abundance of natural gas, by the onset of the 2010s the government had not set any concrete near-term plans for source diversification into cleaner alternatives, such as nuclear or solar. The abundance had also led to an exclusive focus on the supply side, with little done to improve energy efficiency, or cut down subsidies on electricity and water.

Qatar's main domestic source of energy is natural gas (72% of total primary energy consumption in 2010),[59] which is used for electricity generation, desalination and as the petrochemical industry's feedstock. Qatar does not import energy, and the domestic energy mix is completely dominated by domestic oil and gas. Unlike other Gulf monarchies, however, Qatar does not have to use oil for domestic power provision.[60] Generation is primarily taken care of by the Qatar Electricity and Water Company (QEWC), established in 1990; while planning, implementation, transmission and distribution are the responsibility of another company, Kahramaa, which also informally acts as the regulator. Qatar's power and water sector was liberalized in 2000, with the government and QP currently owning two thirds of the QEWC.[61]

Despite the late 2000s economic crisis, Qatar's power demand continued to increase by around 12% per year. The government has managed to keep capacity expansion well in pace with demand: in 2011, a massive independent water and power plant went on stream at Ras Laffan, with a capacity of 2.7 GW of electricity and 280,000m³ of water per day.[62] The government plans to more than double the capacity from 4,480 MW in 2009 to 10,850 MW by 2014.[63] While capacity might not be a problem, domestic availability of gas beyond 2013 is a growing concern, owing to export commitments.[64] In 2009, the shares of domestic consumption were 14% and 24% of total oil and gas production, respectively, still leaving Qatar with an impressive margin in both sectors.[65]

Even though there are no concrete plans in sight for nuclear or solar and the required regulatory frameworks are lacking, evidence suggests that the government regards both as serious mid-term options. In 2012, the QEWC's general manager told the local media of plans to generate 10% of total energy used in the utility sector with solar energy by 2018, while the president of Kahramaa announced that Qatar was 'considering nuclear for peaceful purposes'.[66] Already in 2007, in the wake of rising regional interest, Qatar set up a small National Centre for Nuclear Information under its environmental authority.[67] Suppliers like France and the US have also been eager to promote the option for Qatar. A leaked US embassy cable from 2008 suggests that the Qatar government was at the time exploring the economic potential of the option, particularly given that its neighbour Abu Dhabi was already pushing forward with plans, and Qatar Foundation had expressed interest in developing a nuclear engineering programme for nationals at Texas A&M University's Doha branch.[68] Kahramaa's unpublished 30-year demand forecast from 2008, prepared by a Dutch consultancy, included both nuclear and solar options. The utility has implied that it would choose either nuclear (in the case of high demand) or solar, but not both, while all future plans will be accompanied by expansions in gas-fuelled electricity capacity.[69] In 2009–10 a solar desalination study, commissioned by the newly-established Qatar National Food Security Programme and Kahramaa and conducted by the French company Sogreah, examined the feasibility of a solar CSP plant that would produce a massive 365 mcm of water per year. Also, Kahramaa and the Ministry of Environment are reported to have studied the potential of unilateral and multilateral nuclear programmes.[70] While there is no urgent need for additional energy sources, solar energy in particular is perceived as important for source diversification, saving gas for export, and reducing emissions, but only 'once technologies become cost efficient', as noted in Qatar's current five-year development strategy.[71]

Although Qatar does not have a major domestic energy demand challenge at hand, as a sign of the impact of growing domestic consumption on future exports, gas blocks originally reserved for a cancelled GTL project were reassigned in 2007 to a domestic gas project.[72] Qatar, mainly because of its large energy and heavy industries, boasts of the world's highest per capita energy consumption rates.[73] In 2006, according to QP, the oil and gas industry, flaring, and petrochemical sector together consumed 69% of total energy usage.[74] Nevertheless, consumption patterns of a large segment of residential consumers are extremely elevated too: in 2008, some estimates placed Qatar's electricity consumption per capita as the world's ninth highest—a figure considerably flattened by the large migrant worker segment.[75]

The first desalination plant was commissioned in 1953. Demand has massively increased since, over fivefold between 1995 and 2011 when total annual capacity reached 539 mcm (roughly a third of ADWEA's capacity in Abu Dhabi). In the late 2000s, Qatar made large investments in new production capacity and technologies: in 2009 alone, water and power network expansions and upgrades totalled US$1.9bn. Although the fast population growth has narrowed the gap between capacity and demand, and the government projects a 5.4–7% consumption growth through 2020, existing capacity and planned expansions are expected to suffice well into the current decade even without demand side measures.[76]

Despite the extreme scarcity of conventional water resources, cheap energy has also granted Qatar the questionable honour of one of the world's highest per capita consumption rates: 675 l/d in 2009.[77] Reflecting the large wealth disparities and consumption patterns among the different segments of the population, according to Qatar's General Secretariat for Development Planning, Qataris' per capita consumption was 1,200 l/d and that of non-Qataris only 150 l/d.[78]

As in the other GCC states, subsidies are arguably the most important hindrance to natural resource conservation in Qatar, in addition to creating important financial and opportunity costs. Electricity and water are supplied to Qatari nationals free of charge and sold to industries, non-Qataris and commerce at heavily subsidized prices; QR0.07–0.14/ kwh (US$0.02–0.04, up to twelve times less than the 2011 consumer price in Finland, for example) and QR4.4–5.2/m³ (US$1.2–1.4). In 2003, the government paid QR829m (US$228m) in water subsidies, 63% of which went to Qataris.[79] In 2008, domestic use (non-industrial) accounted for 75% of all electricity usage.[80] Officially, at least, domestic utility subsidies are not a fiscal burden: the share of water and electricity in the government's total expenditure in 2007/08 was reported to be only 1.7% (US$450m).[81] However, the government still only recovers less than a third of the cost of water production. On the supply side, infrastructure is a major source of wastage: leakages in the desalinated water distribution system alone cost the government up to US$275m per year.[82]

Because of the cheap petrol prices (US$0.23–0.27/l in 2012), weak public transport networks, and an unsafe traffic culture, the prevalence and use of large gas-guzzling cars is high, especially among Qataris and wealthier expatriates. Even though a large part of its population still only commutes in crammed buses, Qatar boasts the highest passenger car density rates outside the OECD: 429 per 1,000 residents in 2008.[83]

It is no surprise that Qatar has tremendous room for energy efficiency improvements: a study from 2008 by the United Nations regional council ESCWA concluded that possible savings amounted to 19% in total fuel consumption, 22% in summer peak load, and 1.3 Mt in CO_2 emissions.[84] Efforts to curb consumption through regulation have so far also been half-hearted: in 2008, a law was passed imposing a QR1,000–10,000 (US$275–2,750) fine for using tap water to wash cars and yards, but local newspaper sources suggest that enforcement is mainly symbolic and intended for awareness-raising.[85] As a promising sign of gradual changes to come, coinciding with the 2012 Earth Day, Kahramaa launched a five-year conservation campaign, Tarsheed (Arabic for conservation), which includes a visible public campaign that aims to increase efficiency, decrease per capita consumption, and 'stop waste and irrational consumption'.[86] Despite embracing a plethora of policy and practical measures, the impact of Kahramaa's campaign will arguably remain limited due to its inability to use tariff hikes to change behavioural patterns.

Qatar's regional gas exports

Qatar has since 2007 exported gas through the Dolphin pipeline, which, owing to its low financial returns, can also be seen as a political or even a prestige project. The Dolphin project emerged from the vestiges of a GCC-wide gas pipeline plan, which was initially proposed by Qatar in 1989, and as interpreted by Dargin is used by Qatar to seek to raise its regional status.[87] Under a 25-year supply agreement, as of 2008, 56.6 mcm/d of gas are exported via the Dolphin from Qatar's North Field to the UAE, and 5.7 mcm/d are exported again to Oman, for a price of US$1.25–1.30/Mbtu. Although Dolphin exports in 2011 represented almost a fifth of Qatar's gas exports, their annual value is less than US$1bn/year. By comparison, in 2007, international spot prices fluctuated between US$6 and US$10/Mbtu, and in 2010–11 were mostly at a low of US$3–5/Mbtu. Developing Abu Dhabi's Shah sour gas project is estimated to cost US$5/Mbtu.[88] Not only does Qatar therefore incur an important opportunity cost from selling gas on the regional market instead of exporting it as LNG to destinations in Asia and Europe, it also indirectly subsidizes the UAE's industrialization and Oman's oil production (in the form of enhanced oil recovery), while allowing these countries to continue exporting LNG at profitable prices. This, together with the North Field moratorium and the rising domestic demand said to be taking 'increasing precedence in any future gas allocation', means that Qatar is not willing to increase its exports through the Dolphin pipe-line to its maximum capacity of 90.6 mcm/d, despite calls by the UAE.[89]

As another partial consequence of the limited supply prospects through Dolphin, gas-scarce Dubai has built an LNG import terminal. Summer peak supplies from Qatar, under agreements dating from 2008, commenced in 2010.[90] In 2009 Qatar and Kuwait were in talks on LNG exports, but disagreements on the terms of the contract led to Kuwait signing a deal with Shell instead.[91] In addition to piped and shipped gas, the GCC Interconnection Power Grid, which in 2011 connected five GCC member states (all except Oman), will now enable Qatar to sell any existing surplus electricity to its neighbours. Given, however, what El-Katiri has termed 'peak load collectivism', the concurrence of peak demand in the GCC, regional power trade has still to prove it can serve beyond its current function as an emergency mechanism:[92] in 2009, with four GCC states already interconnected, Qatar refused to export power to Kuwait, on the grounds of a lack of surplus, despite an original deal of 500 MW. The same recurred in 2010, but during a short period that same year, Qatar sold Bahrain 150 MW for two hours per day, giving cautious hope of more substantial exchanges to come.[93]

Human agency: source of fragmented natural sustainability responses

Qatar is ruled by the Al Thani, a dynastic monarchy which, despite internal strife, has come to gain an exceptionally strong hold on power in the state's institutions. Since 1868, the country's and the family's histories have been inseparable.[94] However, owing to two coups during the country's short independence,[95] a national patriarchal figure similar to Sheikh Zayed of the UAE has not emerged in Qatar. Nor is there a clear high-level patron of the environment among the contemporary Qatari elite; this is partly explainable by the concentration of power in the hands of a handful of people, none of whom has taken strong personal interest in the environmental agenda. Qatar's less systematic approach to natural sustainability is, arguably, also due to the locking of most key elite members into already clearly defined areas of personal patronage and external branding. The consequences of this for natural sustainability-related governance, as will be shown below, have been significant, as by the turn of the 2000s this area still remained on the margins of Qatar's political agenda.

Since 1995, Qatar has been ruled by the modernizing autocrat Sheikh Hamad bin Khalifa Al Thani.[96] Sheikh Hamad, who had already been running the country's day-to-day affairs for some years before, deposed his father, Sheikh Khalifa bin Hamad Al Thani, in a bloodless and widely supported palace coup. Previously, Sheikh Hamad had served as commander-in-chief of the armed forces (1972), Heir Apparent and Minister of Defence (1977), Prime Minister (1978) and chairman of the newly-established Supreme Planning Council, in charge of economic and social policies (1989).[97] After surviving at least one serious counter-coup attempt in 1996, he slowly consolidated his power during the late 1990s and 2000s through economic development and promises of political liberalization. In addition to retaining his military posts, Sheikh Hamad commands ultimate loyalty from the entire security establishment, although much of the *de facto* power has now been passed on to his son Sheikh Tamim.[98]

The country's top governance in the 2000s resembled a Roman temple, with Sheikh Hamad at the top, and directly underneath him three or four persons: the Heir Apparent, Sheikh Tamim bin Hamad Al Thani; the Prime Minister, Sheikh Hamad bin Jassim Al Thani; the Emir's wife, Sheikha Mozah bint Nasser; and the Energy and Industry Minister, Abdullah bin Hamad al-Attiyah. In early 2011, this constellation was somewhat shaken by the appointment of a new Energy and Industry Minister, Mohammed bin Saleh al-Sada, in a move arousing plenty of speculation, seen by most observers as a loss of power by Abdullah al-Attiyah, who was temporarily appointed as the head of the Emiri Diwan.

Sheikh Tamim bin Hamad, Heir Apparent since 2003 when his older brother renounced the title, is Sheikh Hamad bin Khalifa's fourth son. He does not hold a ministerial portfolio but has been given an increasing, albeit still limited, role in government and holds a number of other high-level posts, being chairman of the Qatar Investment Authority and deputy commander-in-chief of the Armed Forces, and having oversight of the dynamic General Secretariat of Development Planning, which reports directly to him. Until 2008, he was also the chairman of the Supreme Council for the Environment and Natural Reserves.[99] Furthermore, Sheikh Tamim is responsible for an increasing part of the security agenda and some 'Emiri duties', such as issuing decrees, and state strategy, including launching the Qatar National Vision 2030 planning document. As the patron of Qatar's National Food Security Programme, he has also been pushing food security and renewables onto the domestic agenda.

A distant cousin and close ally of the Emir since the late 1980s, Sheikh Hamad bin Jassim Al Thani has had a long career as the head of different high-level state institutions and ministries. He has been Foreign Minister since 1992 and Prime Minister since 2007.[100] As the front figure of Qatar's dynamic and independent foreign policy, Sheikh Hamad bin Jassim has attracted criticism from other Arab states, especially Saudi Arabia, by his style and outspokenness.[101] In addition, he is the CEO of the QIA and plays a significant role in Qatar's investment, business, property and real estate sectors, the Al Jazeera television network and Qatar Airways.[102] Owing to yet another successful diplomatic *tour de force* by Sheikh Hamad bin Jassim's Foreign Ministry, Qatar, at the time of writing, was preparing to host the 18th Conference of the Parties to the UNFCCC.

Sheikha Mozah bint Nasser is the Emir's second wife, but Qatar's 'First Lady'.[103] She has taken a prominent role alongside her husband in leading the liberalization and modernization of Qatar, and has profiled herself especially through involvement in the areas of education, culture and health. Sheikha Mozah is the chairperson of the Qatar Foundation (QF), president of the Supreme Council for Family Affairs and vice president of the Supreme Education Council, in addition to a number of other related domestic and international titles.[104] Her office holds no constitutional position, and there are consequently no solid guarantees for the next Emir's wife to play a similar role in the affairs of Qatar, nor for the continuation of the Qatar Foundation vision, although the latter is likely to be carried on by one of her sons and daughters, possibly Sheikh Tamim.[105] Alongside promoting research and development into alternative energies and clean tech, she has been promoting sustainable building and design, including through the QF subsidiary Msheireb Properties (est. 2007).

Abdullah bin Hamad al-Attiyah, coming from one of Qatar's most powerful families, is a close adviser and ally of the Emir and served as Energy and Industry Minister and chairman of Qatar Petroleum from 1992 until January and February 2011, respectively, when he was replaced by Mohammed al-Sada. Since 2007, he has also served as Deputy Prime Minister, a post which he in 2011 shared with Ahmed bin Abdullah al-Mahmoud, before the latter fully taking over the post.[106] Alongside Emir

Hamad, al-Attiyah, who also has a reputation for his outspokenness, was the mastermind of Qatar's LNG programme.[107] Minister Mohammed al-Sada, who previously served as Minister of State for Energy and Industrial Affairs (a new post created for him in 2007), holds a degree in marine science and geology and a PhD in engineering, and is said to have inherited a 'fascination with marine life from his father, who was a pearl diver'.[108] Although little is still known by experts of al-Attiyah's technocrat successor's preferences and ambitions, it is possible that renewables and other alternative sources will figure more prominently on the younger al-Sada's agenda.

Other important personalities include the Finance Minister Yousef Hussain Kamal, also in operational charge of the QIA and chairman of RasGas; Ibrahim Ibrahim, a long-time economic adviser to the Emir and the secretary general of the General Secretariat for Development Planning (GSDP) in 2006–11, and his successor Saleh Mohammad al-Nabit; Sheikh Hamad bin Jabor bin Jassim Al Thani, director of the GSDP and Prime Minister Sheikh Hamad bin Jassim's nephew; and the Minister of State for Foreign Affairs since 2011, Khalid al-Attiyah.

Institutions and decision-making dynamics: the broader picture

Despite the constitution of 2005, stating that the country's 'political system is democratic' (Art. 1), Qatar is an absolute monarchy, with the Emir holding *de facto* both legislative and executive powers and exerting wide powers on the direction and pace of development, modernization and liberalization in the country. Kamrava has described it as a 'largely benign autocratic' system, characterized by persisting 'shaykhly patterns of rule', evident in the personalization and concentration of power, the importance of networks of patronage, and opacity of decision-making.[109] Qatar's Council of Ministers, chaired by the Prime Minister, is the supreme executive authority and takes care of the functioning of the ministries and administration generally. The Emir appoints and dismisses the ministers and ratifies all draft laws and decrees drawn up by the Council of Ministers.[110] Ministerial posts, and particularly important ones, are mainly occupied by Al Thani members, altogether eight out of twenty being of that family in 2008.[111] Political parties and trade unions are not allowed, and even the establishment of apolitical professional associations and societies is greatly restricted. The Advisory Council (*Majlis al-Shura*), established in 1972, consists of thirty-five appointed members, generally notables, and has no legislative powers.[112] A permanent constitution, which entered into force in 2005, stipulates the creation of a 45-seat Advisory Council that will have two thirds of its members elected by direct elections and the rest by the Emir, but still remains considerably limited in terms of independence and efficacy.[113] The promises of parliamentary elections, given since 1998, have been postponed several times, most recently in 2010, to 2013.[114]

In addition to consolidating the inheritance of the country's rule among his male descendants in the constitution (Art. 8), Sheikh Hamad bin Khalifa set up the Council of the Ruling Family in 2000. Chaired and appointed by the Emir, the Council decides on the salaries of the ruling family and when power—in case of death or disability—is transferred from the incumbent Emir to the Heir Apparent (Art. 15).[115] In addition to this 'neopatrimonial structure building', new posts and parallel state institutions created over the past decade and a half and staffed by the Emir's allies and close relatives (Qatar Foundation being the prime example) have also been seen as part of the ruler's survival strategy.[116]

Unlike in the UAE, where natural resources are controlled by the individual emirates and the federal Ministry of Energy is weak, Qatar's equivalent is a powerful institution. Until 2011, the Ministry of Energy and Industry was led by one of Qatar's top figures, Abdullah al-Attiyah, who simultaneously chaired Qatar Petroleum and represented Qatar in OPEC's meetings and other international energy fora since as early as 1972.[117] The Ministry of Environment, as in the UAE, is a young and weak institution, with a young and uninfluential minister, Sheikh Abdullah bin Mubarak al-Midhadi, and consequently little sway over other ministries. These relative powers and the (lack of) interplay of the two ministries do not just indicate the importance that Qatar's top ruling elite bestow on energy and heavy industries, even at the expense of environmental sustainability. It is also at the root of the explanation for Qatar's policies and negotiating positions in the context of the international climate regime up until 2012.

In the late 2000s, the leadership embarked on an attempt to streamline and modernize Qatar's government institutions. Seven new ministries were created in 2008 alone, including the ministries of Environment, Social Affairs and Labour. The old and new institutions also function as an important channel of rent allocation by providing a large number of Qataris with generously paid employment. As a result, in 2011, despite the reform, small Qatar sustained a total of thirteen ministries, five councils, seven authorities, eleven other government agencies, and nine other state-owned organizations like the Olympic Committee and Qatar University.[118]

Qatar National Vision 2030

The Qatari government did not miss out on the subregional trend of long-term development planning in the late 2000s. One of the new institutions, the General Secretariat for Development Planning (GSDP) established in 2006, quickly grew into an influential entity through its work in this area, led by Sheikh Hamad bin Jabor Al Thani (Prime Minister Hamad bin Jassim's nephew), as director general, and the Syrian-born Ibrahim Ibrahim as secretary general. In a reshuffle in early 2011, in a possible consolidation of Sheikh Tamim's power base, the former post was replaced in accordance with an Emiri decree by a Supreme Committee for Development Planning, chaired by Sheikh Tamim, while the latter post was passed on to the Qatari academic and economist Saleh Mohammad al-Nabit.[119]

Shortly before the GSDP's establishment, its predecessor, Qatar's Planning Council, had started envisaging a strategic umbrella for the different government agencies' projects.[120] The government had realized that Qatar's development was unsustainable and was placing the country in many ways 'under stress'.[121] Out of this perceived need to simultaneously control growth, diversify the economy, and develop a national knowledge economy, mandated by the Emir himself, the GSDP drafted Qatar National Vision 2030, the country's first major strategic long-term planning document and its first sustainable development plan.[122] Somewhat unlike Abu Dhabi's planning documents, Qatar's Vision, published in 2008, comes across as primarily a document written for and targeted at Qataris. It is riddled with contemporary GCC buzzwords like knowledge-based economy, investing in education and health, Qatarization, national values, wise investment of fossil fuel revenues, fostering entrepreneurship and innovation, and intergenerational justice. Faithful also to global trends and discourse, the Vision is based on the three pillars of sustainable development and includes the notion of human development.[123] Although the GSDP officially has only an advisory role, the Vision has the support of the highest authorities; it was brought to the government by the Heir Apparent Sheikh Tamim and approved by the Emir.[124] Initiated in 2006, the strategy process included consultations with all major national stakeholders, such as ministries and major hydrocarbon companies. Also, the National Development Strategy process included stakeholder consultations. In the spirit of Qatar's new constitution, the Vision stresses the ownership of the document by the Qatari people. Promises of continuing consultations with all stakeholders, including civil society, have also been given.[125] The Vision document as a whole is largely descriptive in style and lacks quantitative goals. Its main purpose is to provide a basis for five-year National Development Strategies, the first of which, consisting of fourteen sector strategies with specific actions and targets, is for the period 2011–16.[126]

Qatar's environmental and climate governance

Environmental governance in Qatar is relatively new in terms of institutions and continues to figure low on the government agenda, always ultimately preceded by considerations of economic development in the energy, infrastructure and construction sectors. A clear institutional framework for coordinated climate change-related governance and policy-making is only starting to take shape. Furthermore, prospects for quick improvements are doubtful, as there are no powerful elite personalities identified as leading either agenda. The continued dominance of the energy and industrial sectors over environmental issues in their respective spheres, and the weakness and low level of performance of the new Ministry of Environment, have hindered the development of an effective, competent and overarching environmental authority. As another result of this, top-down efforts to integrate natural sustainability and climate change considerations in cross-sectoral policy-making have been largely weak in implementation, with the most notable efforts coming in the form of medium-term planning from the GSDP and the food security programme QNFSP.

Despite the lack of a clear leader figure in environmental sustainability, interviews with local and Qatar-based stakeholders confirm that a number of persons and institutions are domestically associated with 'green qualities': Emir Hamad and his wife Sheikha Mozah are regarded as the driving forces behind the Qatar National Vision and behind the Qatar Foundation's energy- and environment-related research developments, respectively. The Ministry of Environment and Minister Abdullah bin Mubarak al-Midhadi, and the Ministry of Energy and Industry/Qatar Petroleum and Minister Abdullah al-Attiyah, because of their respective roles, have also been seen as important. Finally, Heir Apparent Sheikh Tamim, as the patron of Qatar's food security programme, is perceived to have a role to play in the areas of renewable energy, water management and sustainable agriculture.[127] Also, 'lower-ranking' local environmental (sustainability) pioneers include Saif al-Hajari, the chairman of the Friends of the Environment Centre and Qatar Foundation's vice chairman until 2012; Issa al-Mohannadi, chairman of the Qatar Green Building Council and CEO of the QF subsidiary Msheireb Properties; and Yousef al-Horr, chairman of the Gulf Organisation for Research & Development, which has developed a local sustainable building rating system.

During the past three decades, Qatar's official environmental governance structures have been gradually upgraded, but their capacity has always lagged behind the country's fast development, leaving them insufficiently empowered. The Permanent Environment Protection Committee (PEPC), established in 1981, was replaced in 1994 by an environmental department established under the Ministry of Municipal Affairs and Agriculture in 1994, the tasks of which were to monitor and mitigate pollution and conduct related studies and environmental impact assessments. At the same time, records also exist of an Environmental Technical Committee (ETC), comprising all major industrial companies.[128] The tasks of the Environment Department were transferred in 2000 to the Supreme Council for the Environment and Natural Reserves (SCENR), whose tasks were in turn taken over in 2008 by the newly-established Ministry of Environment.[129] The PEPC and the ETC, both under the Ministry of Municipalities and Agriculture, served mainly as discussion forums for legislation and regulations. A UN report from 1999 laments that despite many promising projects and programmes, the PEPC lacked financial and human resources for their implementation.[130] The competences of the SCENR were considerably extended from this, and they officially included the formulation of policies, formulation and enforcement of legislation, monitoring and mitigation of pollution, environmental impact assessments, wildlife protection and monitoring activities, awareness raising, and operation of a national environment database.[131] The Ministry of Environment, established in 2008, covers the same main functions as the SCENR, including the responsibility to protect the environment, endangered wildlife and natural habitats, but also has many new task areas.[132]

Before its replacement, the SCENR had a staff of 600, half of whom were involved in surveillance. According to Richer, the Council's capacity was not sufficient, and it was not able to keep pace with the large number of environmental impact assessments resulting from the increasing development activities: over 1,000 in both 2006 and 2007.[133] Partly, perhaps, because of the SCENR's short existence, a report by the GSDP in 2009 still described it as 'under-resourced in terms of staff and expertise'.[134] By the end of 2010, the Ministry's total staff had grown to a massive 2,700, of whom 1,400 were Qataris.[135] While the problem of quantity had finally been solved, other fundamental problems still persisted, as in the same year stakeholders characterized the Ministry as being reactive rather than proactive, not working '100%', having an extremely low responsiveness, and even working less effectively than the SCENR.[136] The weak power base of the new Environment Minister, Abdullah al-Midhadi further reduces the prominence of his young ministry in Qatar's personalized institutional hierarchy.

In addition to the ministry, and because of the vacuum formed by its lack of human capacity, a handful of other institutions engage in environmental governance: the GSDP's work with the National Vision and the development strategies encompasses environmental development, also discussed further below. Qatar Petroleum's Health Safety and Environment (HSE) department works with environmental conservation, sustainable development and oil spill response preparedness, among other things, relating to the company's operations. QP also has a sustainable development department.[137] Furthermore, the industrial cities of Mesaieed and Ras Laffan, under QP, as well as the land reclamation real estate project The Pearl, developed by UDC, function as 'mini states within a state', conducting their own environmental planning, monitoring, and even regulation. For example, the Mesaieed Industrial City Authority, established in 1996, acts as the single point authority and issues environmental guidelines and protection criteria for industries operating in the area.[138]

Environmental regulation, planning and policies: theory and practice

Despite the gradual build-up of an environmental institutional and regulatory framework since the 1980s and into the 2000s, by the end of the decade implementation still remained incomplete. Most likely Qatar's first law in the area of environment was law no. 4/1981, which established the Permanent Environment Protection Committee.[139] In 2002, an environmental law of 75 articles was enacted.[140] Other laws regulate hunting of birds and reptiles (law 4/2002), protection from radiation (31/2002), protection of wild fauna and flora and their natural habitats (19/2004), and environmental protection in general (executive by-law 4/2005). There have also been Emiri decisions calling for the protection of the marine environment, and the permanent constitution of 2005 lays down that 'the State shall preserve the environment and its natural balance in order to achieve comprehensive and sustainable development for all generations'.[141] Environmental impact assessments (EIA) have been carried out in Qatar since the 1980s and were made compulsory for all development projects, including industrial projects, in 2002.[142] In practice, however, serious problems in their implementation persisted at the turn of the 2010s; observers pointed to weak enforcement, despite guidelines having existed for fifteen years. The number of evaluation personnel at the Ministry of Environment is highly insufficient, and set criteria for requirements and 'quality control' are said to be lacking. In 2007, of more than 1,000 projects that were submitted to the SCENR, only 5% failed to pass scrutiny.[143]

Indeed, the existence of regulatory mechanisms has not implied their full enforcement. Apart from the environmental laws and the EIAs, the Ministry of Environment seems to have few other regulatory tools at hand. The state currently appears to offer no economic incentives or disincentives to less polluting or emitting, or more energy or water efficient, industries or buildings. Neither are there any major state-sponsored environmental prizes. According to the GSDP's own assessment, while regulatory and managerial bodies have been established quickly and Qatar has joined a number of international treaties in the area of sustainable development, there are still 'numerous challenges' in 'putting theory into practice'. Both internal and external evaluations have pointed out a number of problems and their symptoms: limitations in institutional and human capacity and gaps in data and research are a major constraint to institutional effectiveness and implementation of sustainable development plans. The fast speed of development activities has left the regulatory and monitoring bodies incapable of fulfilling their functions. Richer has also observed a mismatch between offsetting activities assigned to industries (including research and conservation programmes) and the actual damage inflicted on the environment.[144]

An important point, indicative of the paramount importance of the oil and gas industry to the country's economy, is that Qatar Petroleum, like ADNOC, self-monitors and enforces environmental performance criteria in the entire hydrocarbons sector, including its own operations and the industrial cities.[145] The company's HSE Regulation Authority is responsible for norm and standard development and harmonization in the hydrocarbon sector. It also audits the industry in areas such as air quality.[146] A further factor limiting the already narrow scope of action of the Ministry of Environment is that a fourth of Qatar's factories were built before 2005 and fall outside existing legislation, meaning that they need not follow environmental standards until the end of a transitional phase, in 2014.[147] Sadly, older industrial installations are almost invariably the most inefficient and harmful to the environment.

All this implies that, although a number of departments were moved under the new Ministry of Environment from other ministries[148] and it has otherwise grown in staff and rank, it seems still to have little or no ability to keep checks on the hydrocarbon and industrial sectors, which—as in the other Gulf monarchies—are the largest sources of pollution, emissions and environmental degradation. An external scientific report in 2004 found that the best areas for coral growth, located around QP's main oil and gas terminal island, Halul, were suffering from 'significant human impact' from marine outfalls and dredging, among other causes. Very high coral mortality from bleaching and human impact since the mid-1990s has been reported especially in areas close to the coast.[149] Also, the carbon footprint of the hydrocarbon industry is massive: around 70% of Qatar's greenhouse gas emissions in 2006.[150]

Indeed, the available evidence strongly indicates that industrial and economic growth and the corresponding state institutions still weigh more in state decision-making than environmental concerns and corresponding authorities. While it is certainly part of Qatar's state-building, institutional reform and perhaps even external image-raising processes, the Ministry of Environment still has a long road ahead to establishing itself as the main environmental institution of the state.

While Abu Dhabi's responses to natural unsustainability are (or will be) dispersed among a few key thematic and sectoral planning documents, Qatar's approach in this particular sense has been strictly top-down. All subsequent sectoral development plans are supposed to follow from the broad Qatar National Vision 2030 document and the more detailed and goal-oriented five-year development strategies derived from it. Regarding the environment, the Vision itself recognizes the adverse environmental effects of Qatar's recent fast and partly uncontrolled growth, but fails to set these ahead of the development imperative: it admits that 'even with Qatar's best efforts, it is impossible to entirely avoid harming the environment, given a development pattern that depends in its early stages on oil, gas, petrochemicals and heavy industries'. The document establishes that neither economic development nor protection of the environment should be sacrificed for the sake of the other, and offers advanced technologies and avoidance of rapid, unplanned growth as the solutions to this dilemma. It also formulates three climate change-related goals: encouraging regional cooperation in preventing and mitigating development activity-related pollution; taking a 'proactive and significant regional role' in climate change impact assessment and mitigation; and supporting international climate change mitigation efforts.[151]

The operational document for 2011–16, the first National Development Strategy, outlines the areas of focus in the field of natural sustainability; among Qatar's '20 key challenges', it lists low economic diversification, inefficient use of resources (including natural), unsustainable water consumption patterns, and environmental stress from urbanization and consumption patterns. As responses to these challenges, and to promote 'sustainable prosperity', the document prescribes 'a programme of strengthened environmental management across economic and natural resource sectors', including an integrated water management plan, a solid waste management plan, stronger demand management for power and fuel, agricultural reforms, and promoting sustainable urbanization. The document also stresses the concept of intergenerational justice as the basis for environmental actions.[152] As in the UAE and Saudi Arabia, sustainable building has become a buzzword in Qatar's booming construction sector. No compulsory codes or ratings, however, had been implemented by 2012.

Climate policy decision-making structures

A national committee on climate change has officially existed since 2007, first under the SCENR and then under the Ministry of Environment.

Despite this—and some other lower-level UNFCCC-related commit-tees and points of contact and authority—by the end of 2011 there had been no major visible institutional or policy evolution in relation to the climate change issue in Qatar. While in Abu Dhabi non-oil sector insti-tutions have increasingly taken over the climate agenda, in Qatar the absence of an environmentalist patron-figure and the powerful 'old energy' sector's lack of interest in alternative energies and energy effi-ciency had shaped the country's international policy response into a defensive rather than proactive one. This also hindered domestic level developments in this sphere. A sea change, however, is likely to take place in 2012, as will be discussed in chapter 8: in November 2011, contrary to some expectations, Qatar won a bid against South Korea to host the 18th Conference of the Parties (COP18) to the UN Framework Con-vention on Climate Change convention (UNFCCC) in 2012. Although close observers have confirmed that the Ministry of Foreign Affairs, lead-ing the campaign, had originally perceived COP18 as little more than yet another conference in a series of many already hosted, and thus in line with this niche of Qatar's external branding strategy, a profound rethink and definition of Qatar's external and domestic climate policies were imminent as international attention began turning to the fossil fuel-exporting host country with a dubious negotiation reputation and few domestic actions to claim 'climate credibility'.

Overlooking bad press, slacking awareness-raising: perceptions and policy-making

Having developed a thick skin from its maverick foreign policy and dip-lomatic ventures in the MENA region in the 2000s, the Qatari govern-ment has received external criticism regarding its environmental unsustainability and greenhouse gas emissions in a much more relaxed way than its south-eastern neighbours Dubai and Abu Dhabi, tending towards defensiveness more than direct actions to improve its interna-tional image. The task of evaluating Qatar's relative environmental (un)sustainability, however, is complicated by lack of up-to-date information and important gaps in data in many areas. On account of Qatar's weak tradition of statistical data gathering and aggregation, and its small pop-ulation, only a few comparative data sets of the country's current relative environmental performance exist.[153] In the Yale and Columbia Environmental Performance Index of 2010 it fared almost as poorly as its GCC peers, ranking as the 122nd(/163), ahead of Oman (131), Bahrain (145) and the UAE (152).[154] In the same year, the WWF's Ecological Footprint Index ranked Qatar, included in the study for the first time, as second worst in the world (10.5 global hectares per capita), almost side by side with the UAE (10.7 gha).[155] Two years later, Qatar had climbed to the top of the list.

Qatar's second Human Development Report, published by the GSDP in 2009, is a relatively frank portrayal of the country's environmental problems and challenges and how these stem from the uncontrolled and rapid growth. The report calls for a 'change in mindset and in consumption and production patterns' and recommends the drafting of a sustainable development policy framework that would cover the issues of climate change (emission management and incentives for industries) and water and energy use (including investments in new technologies, such as renewables).[156]

In the late 2000s, general awareness of environmental sustainability and climate change in Qatar was still very low. At the government level the Ministry of Environment is tasked with awareness-raising and education in environmental issues, but its work and activities have so far been rather limited in scope, visibility and quantity.[157] In addition to being the front-runners in environmental protection,[158] the energy industry and private companies have been slightly more active in terms of organizing one day campaigns, training events, and seminars on a range of topics. However, action remains low-impact and lacks consistency. Greenwash is common, taking the form of beach clean-ups, tree planting events and other one-day awareness campaigns, seminars, conferences, competitions and awards. Also, because of the sheer size of many of Qatar's industries, even significant improvements in efficiency and environmental performance can easily appear inconsequential: Qatalum's aluminium smelter, inaugurated in 2010 and dubbed one of world's most efficient and environmentally friendly smelters, still uses up to 1,350 MW of state-subsidized, greenhouse gas-emitting, fossil energy (equal to the capacity of a large nuclear reactor).[159]

Until Kahramaa's conservation campaign in 2012, top-down awareness raising by official authorities had been largely invisible to the general public, and bottom-up actions of companies and associations remain naïve or self-limiting and avoid sensitive topics, such as industrial pollution or the unsustainability of the local lifestyles. Furthermore, Qatar's awareness-raising efforts, like those of the other small Gulf monarchies, are complicated by the large and diverse expatriate community with often low levels of awareness to begin with and no channels of participation and, as a consequence, little sense of accountability.

There are a handful of NGO-like environmental organizations, formed by either nationals or Western expatriates, which only manage to reach small fragments of the society. Furthermore, all Qatari-led organizations have been co-opted by the state in one form or another. The Friends of the Environment Centre, founded in 1992 and led by Saif al-Hajari, works largely with children and youth, through camps, awards and thematic programmes, including the annual Flower Each Spring campaign.[160] The Qatar Green Building Council (est. 2009) is another semi-private organization, established by Issa al-Mohannadi, the CEO of the Qatar Foundation subsidiary Msheireb Properties.[161] Qatar Green Center (est. 2005), under the Ministry of Environment and led by Abdullah bin Mohammed al-Kuwari, concentrates on educating Arabic-speaking children in tree-planting, landscaping, recycling and litter reduction, among other things.[162] Among the few non-Qatari environmental NGOs are the largely Western expatriate Qatar Natural History Group (1978), mainly focused on arranging monthly lectures and fieldtrips, and the more recent and smaller SustainableQatar (2008), which largely targets professionals and concentrates more on professional networking and lectures, and has recently managed to engage Qatari participants as well. The local UNESCO office (early 1970s) has also organized some activities in the areas of recycling and environmental awareness among local schools.[163] Local dignitaries have also made financial contributions for conservation, including a US$1m donation in 2008 from Sheikha Jawaher bint Hamad bin Suhaim Al Thani, wife of the Heir Apparent Sheikh Tamim, to Birdlife International, managed by the Friends of the Environment Centre.[164] Prior to 2011, one of the few visible regulatory enforcement efforts on the government's side had been Kahramaa's rather half-hearted campaign in 2010 to enforce law 26/2008 on conservation of water and electricity.[165]

In the absence of reliable polls, anecdotal and observation-based factual evidence suggests that environmental and climate change awareness among Qataris is low. However, young Qataris confirm there is an increasing interest in this, albeit still narrow and unstructured.[166] A value survey by Qatar University from 2010 indicated extremely high identification with a concern for the environment and its protection (98%), but the reliability of the poll's results is called into question because an impossibly high proportion of surveyed Qataris claimed to have participated in an environmental demonstration in the previous two years (12%), although most certainly no such things have been organized in the country.[167] As in Abu Dhabi, expatriates in Qatar either come from countries where environmental awareness is low or are alienated from their everyday 'green' practices by the lack of an infrastructure for sustainability (public transport, recycling bins, saving energy, and so forth). Arguably the often consumerist, car-heavy lifestyles of the highest income/wealth segment of the population, which is a small minority largely composed of Qataris, also have an observable psychological impact on lower income groups. Surrounded by large SUVs, lofty houses and luxury stores, many in latter groups begin seeking consumption patterns and status symbols that resemble those of the former group of 'alpha consumers' for increased social prestige. Where the urban infrastructure and planning do not encourage individual-level natural sustainability, and channels to influence decision-making are minimal for the average person, apathy reigns. And where money and individual career opportunities abound and are often acquired with comparatively low effort, a sense of purpose vis-à-vis what Giddens has termed as 'back-of-the-mind issues',[168] like environmental degradation, becomes diluted.

7

QATAR'S CLIMATE CHANGE
AND SUSTAINABILITY RESPONSES

By the early 2010s, despite its high vulnerability to climate change, relatively little had happened in Qatar in terms of climate change governance, institution-building, and domestic policies and measures. Unlike Abu Dhabi, which had by then successfully branded itself as the clean energy leader of the Gulf and had already begun moving towards implementation in the deployment of alternative energies and technology transfer, Qatar had chosen a more gradual approach. With abundant gas reserves there was little rush, and because of the limited extent of homegrown human resources, the areas of focus needed to be carefully handpicked. Half intentionally, half involuntarily, the climate change, alternative energy and environmental sustainability-related projects and initiatives that began taking form in Qatar in the latter part of the 2000s were clearly distinct from those in Abu Dhabi. At the turn of the decade, with the Qatar Foundation-linked initiatives, including a technology park, representing the most important cluster of related efforts, the contours of Qatar's climate change puzzle were only beginning to appear.

Climate change impacts

The World Bank report of 2007 classifies Qatar as the most vulnerable Arab country and the third most vulnerable developing country in terms of climate change impact on land area from sea-level rise, with projections ranging between 3% and 13%. A consultancy study from 2009 extends the upper range to 18%. Qatar's economy also ranks among the most vulnerable to sea-level rise, with 2–5% of the country's GDP at risk in 1–3-metre scenarios. In the case of an extreme sea-level rise, of 5 metres or so, studies estimate that 10–50% of the country's population would be affected.[1] Generally however, country-specific data, particularly regarding potential societal impacts of climate change on Qatar, is extremely scarce. The state has submitted one national communication to the UNFCCC, but no comprehensive country-specific impact studies are now known to have been undertaken. Existing information on Qatar's historical climatic patterns, however, suggests that the annual mean temperature has increased by 0.3°C since the 1970s, because of a significant increase in night time temperature.[2]

The GSDP's documents, including the National Vision, recognize the multiple challenges of resource scarcity, environmental degradation and climate change, ranging from diminishing water resources, pollution and environmental degradation to the potential impacts of climate change on coastal developments.[3] According to the body, Qatar's main climate change-related risks and vulnerabilities include flooding and loss of land area; damage to the marine environment; water stress; food insecurity; dependence on the oil and gas sector; inability to transform into a low-carbon economy; high cost of long-term adaptation; inappropriate education and training; and health risks.[4] Detailed, numerical studies, however, do not exist, and local experts have called for an assessment of the potential impacts of climate change, adaptation strategies and policies, and high resolution climate change scenarios that would serve a small country like Qatar.[5] Qatar's first and so far only communication to the UNFCCC, published in 2011 by the Ministry of Environment, highlights the 'further desertification of the desert' and increased water demand as the main consequences for Qatar of climate change-induced temperature rise coupled with no increase in rainfall patterns. Increased water demand, along with increased need for air conditioning, will lead to increased energy demand, which produces more emissions and waste heat. Given the shallow waters of the Gulf (with an average depth of 20–30 m around Qatar), threats from climate change to Qatar's marine biodiversity could also be devastating.[6]

No major studies exist concerning the potential economic impacts of climate change on Qatar (as is the case for most states), but there are a number of modelling studies on the potential economic impacts of 'response measures', or international mitigation policies and measures. Despite presenting a mixed picture of Qatar's vulnerability, they generally point towards the country's natural gas reserves as being a strong asset in countering these potential future impacts. Earlier comparative studies that predicted 3–5% GDP losses for Qatar had by the 2010s become outdated,[7] with none of the predicted losses demonstrable. Local authors writing in the 2000s were also convinced of economic impacts to come, although to varying degrees: in 2004, Ahmed and al-Maslamani examined a number of models that estimated the adverse impacts of response measures from the implementation of the Kyoto Protocol for Qatar and found significant variation between models, depending on how implementation is managed by the Annex I (developed) countries. They also noted the high uncertainties involved in predicting future impacts.[8] Al-Mulla, surveying existing studies in 2009, also reached the conclusion that the implementation of the Kyoto Protocol would 'unquestionably' result in adverse economic impacts on Qatar.[9] Evidently, none of these losses were visible by the early 2010s, and as the studies also note, Qatar's gas abundance, its massive export schemes, industrialisation strategy, and the high GDP per capita levels will be important factors in offsetting any future impacts. The global application of market mechanisms (emissions trading schemes) and/or the adoption of carbon taxes are expected to lead to natural gas being preferred over oil and coal.[10]

However, owing to Qatar's dependence on exports of fossil fuels and energy-intensive products, the GSDP for example believes the country's export income will decline in the long term, especially if importing states impose border taxes. The UNFCCC communication too stresses the critical importance of economic diversification 'within the hydrocarbon sector and away from it' for future economic stability.[11] Apart from the significant investments needed for diversification to a low-carbon economy, the GSDP expects climate change adaptation-related infrastructure investments to be 'very high'. It is feared that Qatar's economic situation will significantly deteriorate if declining revenues coincide with the need to invest in adaptation or diversification, which could have significant negative effects on domestic food and water security and human development—not to mention the sustenance of the existing rentier bargain.[12] This same challenge is obviously shared by all GCC states.

Resource scarcities, related vulnerabilities and adaptation measures

At the turn of the 2010s, Qatar is the only Gulf monarchy not suffering from domestic energy supply shortages or expected to face relative domestic energy insecurity in the near future. Nevertheless, like the UAE, Qatar has major future challenges in the areas of water and food security.

In its small territory, Qatar has extremely scarce water resources, which are vastly overexploited. It has some underground aquifers, but no lakes or rivers, and one of the world's lowest precipitation rates (20–150 mm/year).[13] Depending on the source, Qatar has from less than 100 to less than 200 m³ of natural water resources per capita per year, which is clearly below the water poverty line of 1,000 m³ per capita and among the lowest in the Arab world.[14] Overuse of groundwater has been traced back as early as the late 1960s, caused by the increasing number of farms, which was in turn prompted by government investments in the sector.[15] As early as 2002, Qatar's water stress index was 157%, meaning that the country's renewable water reserves were exhausted and non-renewables were being exploited. By the late 2000s, depending on the source, the water abstraction rate was four to six times the natural groundwater recharge rate.[16] Desalination, which has been used since 1954, provides for practically all domestic water usage. Roughly half of all water used is either desalinated seawater or treated wastewater (used for forage irrigation and landscaping). Bottled water, in high demand in households, is imported from at least twenty-three countries.[17]

In a water scarce country, short-term security is a major priority. Around the mid-1990s Qatar's total potable water storage was 1.1 mcm, equal to three days' supply. In 2008, after a massive population increase, despite an estimated storage capacity of 2 mcm, there were only 1.5 days' supply in storage (at normal consumption rates).[18] In 2010, Kahramaa ordered a feasibility study for its Water Security Mega Reservoirs scheme, which envisages a linkage between two desalination plants and five reservoirs, with the aim of providing for one week's emergency supply, extendable to one month's under rationing. Work on the US$2.75bn project is expected to begin in 2013.[19] In the long term, water security could also be compromised if capacity does not follow with demand growth. In the 1990s, water production grew by an average of 7% per year.[20] No time series for water consumption growth for the 2000s were available, but according to a British estimate, summer water consumption increased by

7% from 2008 to 2009.[21] In 2009, Kahramaa estimated that Qatar's water demand would rise by roughly 50% over the next decade.[22]

Supporting domestic agricultural production clearly has more to do with perceived rather than actual food insecurity: in the absence of major international trade dysfunctions, Qatar and the other GCC states have little to worry about. However, prompted by the global food price crisis of 2007–08, Qatar—unlike Saudi Arabia, which is reversing its long-standing and highly unsustainable self-sufficiency policy—is opting for a three-pillar food security strategy, based heavily on boosting domestic production, while also securing exports through trade and overseas agro investments and building a strategic food reserve of staple crops for a 'chaos' scenario.[23] When implemented, the food security strategy will reverse Qatar's long-term water strategy of using groundwater for agriculture and desalinated water for potable supply only.[24]

As a result of long-standing food self-sufficiency policies, agriculture continues to be the major consumer of water in Qatar, taking up 74% of freshwater use while contributing only 1% to GDP. By the 2010s, as a consequence of government incentives for farmers, the number of farms in Qatar had risen to over 1,200 from 453 in 1975, although simultaneously many farms continued to be abandoned on account of groundwater depletion.[25] Since water for irrigation is free and pumping costs are minimal, there are no saving incentives for the farmers.[26] In general, agriculture is not an important source of income for Qataris and the people employed by the sector in 2007, around 16,000, were all non-citizens.[27]

In addition to consuming huge quantities of groundwater, current agricultural patterns create a number of problems for water and food security, including soil salinisation, land degradation and saltwater intrusion into the remaining fossil water resources.[28] As a result of the overexploitation of groundwater, mainly by agriculture, salinity increased from 1971 to the mid-2000s by 67–100%.[29] Moreover, if temperatures rise, evapotranspiration and water demand for agriculture, and areas for grazing that rely on irrigation will increase, creating ever larger stress on water provision.[30] A massive planned agricultural expansion, prompted by the new Qatar National Food Security Programme, including 1,600 new farms by the mid-2020s, would further raise water demand by over a million cubic metres per day, which would inevitably have to come from desalination.[31] Interestingly, the Qatari government, which plans to produce this water with low-carbon energy sources, seems to have found a way to circumvent the seemingly insurmountable challenge of growing food in the desert in a naturally sustainable way—albeit for the time being only on paper.

A key factor contributing to domestic water consumption and wastage is pricing. According to a long-standing government policy, as noted above, Qataris mostly do not pay for water and non-Qataris pay a subsidized price. In the 1950s and 1960s, the government attempted to establish a pricing system in Doha for services, including water, but owing to the poor economic situation of most Qataris at the time, and the belief that water is a gift from God, people refused to pay. Later on, as oil wealth began to accrue, the need to collect payments from the nationals receded.[32] Local studies blame low awareness on the country's extreme water scarcity and the cost of desalination for domestic wastage. The lack of education of foreign servants coming from water rich countries, a lack of educated personnel for efficient water management, and sparse, often solely Arabic-language awareness campaigns are stated as additional causes (which notably place the fault on others than nationals).[33] Qatar has an extremely high per capita water consumption level, despite all potable water being desalinated: 310–675 l/d depending on the source.[34] There are huge disparities between a Qatari's per capita consumption, at 1,200 l/d, and that of an average foreign resident, at 150 l/d (close to the OECD average), presumably due to the generally larger houses and gardens of the former group.[35] Qatar's houses, villas and palaces consume astronomical amounts of water, ranging from 14,000 l/d for a house without a garden to 20–35,000 l/d for each of Qatar's 639 palaces in 2003.[36] Losses from transmission and distribution are also a major contributor to water wastage, stemming from 'corrosive soil, poor installation techniques and maintenance, and improper design'. They have been estimated to amount to a third of total water consumption in 2007–11, while in the OECD, average losses are 18%. Desalinated water leakages also push up Doha's water table, causing health hazards and increases in construction costs.[37]

The GSDP has aptly remarked that 'Qatar's water crisis is essentially a crisis of governance'.[38] Management has always focused on the supply side, with large amounts invested in desalination and, with subsidy cuts apparently ruled out in the near term, efforts will be directed at cutting network losses to 10% and applying more efficient, reverse osmosis technologies in desalination.[39] Plans to formulate a management and development strategy with a vision to 2050 were initiated in the mid-2000s, but are still to yield results.[40] The five-year development strategy is expected to bring improvements in this sense, as it identifies water scarcity as the most pressing environmental concern facing the country. It also announces the formulation of a National Water Act by 2016, which will include quality requirements, discharge controls and conservation incentives. Among the demand side measures, the government is considering introducing similar tap devices to those in Abu Dhabi and installing water meters in the commercial sector. Plans are also to encourage less wasteful garden and agricultural irrigation methods and the use of pool covers.[41] Two GSDP reports from 2009 and 2011 boldly call for a revision of the subsidized water and electricity supply policy, but efforts to implement such a measure have fallen short.[42]

Use of recycled wastewater began in Qatar as early as 1971: two large treatment plants account for most of the output: 140–150,000 m³/d in 2005. While the share of recycled wastewater produced from potable water is proportionally higher than in Abu Dhabi (where a fifth of desalinated water is recycled), it is still only about one-fourth, on account of losses through garden irrigation, car washing, leakages, and industries and suburbs that are not linked to the wastewater system.[43] Nevertheless, because of low demand, less than two thirds of treated sewage effluent are currently used, while a part of the wastewater produced is discharged or stored untreated. The current five-year strategy aims to expand existing distribution networks to increase this share, and envisages using treated wastewater for aquifer recharge to enhance water security.[44] Private sector participation in the wastewater sector, generally thought to increase efficiency, began in 2006.[45] A Singaporean company has secured a deal to build a 439,000 m³/d plant north of Doha by 2020, which is envisaged to be surrounded by a 4x4 km offset-type eco oasis park.[46]

Despite the renewed attempt to increase food self-sufficiency, Qatar is highly dependent (90%) on imports of several key staples: 98% of consumed wheat and rice, for example, is imported.[47] The situation is perceived as risking severe supply problems if climate change advances and prompts a decline in global agricultural production while the world's population continues to grow. Besides the restricted water supply options, increases in domestic agricultural production are significantly limited by Qatar's small territory: a survey from 2005 found that 5% of the total land area is arable. However, in 2008 only a fifth of this was being cultivated.[48] In addition, population growth will further add to the challenges in this area, as the amount of both domestic production needed to achieve higher levels of self-sufficiency and the import bill will grow accordingly: according to some estimates, the value of the country's food imports, which stood at US$1.2bn in 2009, could nearly triple by 2020.[49]

In the 1990s, Qatar launched several projects to enhance its water security by importing water through pipelines from neighbouring countries, including the 'Peace Pipeline Project', but all withered away due to regional instability and domestic security concerns, and technical and economic unfeasibility. For example, in the 1990s, Qatar negotiated over importing irrigation water from Iran's Karun river. Another option studied, without positive results, was the artificial recharging of groundwater aquifers.[50] The issue was taken up again in the late 2000s. With the 2007–08 global food crisis, food security emerged as a new priority on the government agenda,[51] and Qatar became active, along with other Gulf monarchies, in the so-called foreign farmland grab. In 2008, the Qatar Investment Authority established Hassad Food, an agricultural finance company with an initial capital of US$1bn aimed at securing food supplies for the country.[52] Later the same year, the Qatari government announced plans to lease an area of 40,000 hectares in Kenya for fruit and vegetable cultivation, in exchange for a US$2.3bn loan for building a second deep-water port in the host country for its future exports to Qatar. The purchase attracted fierce criticism from local conservation groups, community leaders and international analysts, and it was accused of both threatening local ecological diversity and being morally wrong, given that Kenya is a recipient of international food aid.[53]

In August 2009, following international criticism directed mainly at Gulf food investors, Hassad Food's chairman announced that the company would be investing in stakes in agricultural companies instead of land purchases; the company did not 'want to be in a situation where the rich are taking away food and land of the poor'. Since then, however, the company is reported to have bought 100,000 hectares of land in Sudan.[54] Hassad Food also has a presence in Australia and is developing poultry farms and plans to construct pilot greenhouses at home.[55] In addition, it has explored share acquisitions in Latin American agricultural companies and rice farm purchases in Asian countries, and is planning to establish a US$100m company for sheep rearing and grain cultivation in Turkey.[56]

Furthermore, in late 2009, the government launched the Qatar National Food Security Programme (QNFSP), under the Office of the Heir Apparent Sheikh Tamim, tasked with the preparation of a comprehensive food security policy by 2012 (later pushed back to 2014), which will include measures for both increasing domestic food production and securing import supplies (see below).

As in the case of the UAE, or even more so, Qatar's water and food security challenges in the next decades are evidently more linked with the naturally unsustainable consumption patterns than with climate change impacts. Consequently, government responses, while recognizing the synergies of addressing the issues in rhetoric, are principally geared at these more immediate stress factors. Similarly, despite negative impact scenarios of potential climate change-induced sea-level rise, as of 2012 there was still no evidence of government adaptation plans or measures for safeguarding the country's coastal settlements and infrastructure.

From food security to natural sustainability?

In 2008, prompted by the global food price crisis, the Qatar National Food Security Programme (QNFSP), a sectoral initiative with a seemingly narrow task, to enhance Qatar's food security, began taking shape. It was launched in 2009 under the patronage of the Heir Apparent Sheikh Tamim as a task force to devise a national master plan for food security. Endowed with a massive budget, of an undisclosed size, the QNFSP, led by his dynamic and visionary ally Fahad bin Mohammed al-Attiyah, coordinates an inter-governmental task force of over a dozen institutions and companies. The two main dimensions of the QNFSP's work are sustainable domestic agricultural production and strengthening the security of food imports to Qatar.[57] Regarding the former, the QNFSP's objectives include improving Qatar's food self-sufficiency and furthering economic diversification by implementing a highly ambitious solar desalination scheme. The level of self-sufficiency aimed at by the mid-2020s will vary from staple to staple, and could be anything between 10 and 80%, according to experts working for the task force in 2011. The programme's timeframe consists of a currently ongoing planning phase and an implementation phase, due to begin in 2014. QNFSP's water and energy engineering department, working together with Western consultancies such as the German DLR,[58] is studying the feasibility of several technology options in a number of areas, including desalination, solar and wind energy, aquifer storage of water, and smart grids. The programme is currently evaluating the possibility of building up a desalination capacity of 1 billion cubic metres per year, for the exclusive use of domestic agricultural production. The needed energy is envisaged to be produced from clean sources, most likely solar, but by 2011 the QNFSP had opted for a technologically neutral planning strategy, which leaves more flexibility to incorporate new and enhanced technologies as they become available during the 10-year implementation period. So far, studies commissioned by the programme indicate that solar desalination would become economically feasible in high LNG export price scenarios.[59] Other aims laid out in the QNFSP's core concept, which draws heavily from the National Vision document on the rhetorical side, include becoming 'an exporter of knowledge-based solutions and technologies' by developing R&D expertise in the areas of arid land agricultural production, desalination and water management, renewables and food processing.[60]

In 2012, it is still too early to evaluate the chances of success of the high-aiming programme. As in the case of Abu Dhabi's Masdar City, economic viability considerations, the viscosity of most other government institutions—and arguably, to some extent, scepticism among potential (local) investors towards the planned technologies—are likely to form a barrier between some of the plans and actual implementation. Furthermore, with a current population of 1.7 million and high economic and population growth rates, the programme's self-sufficiency goals might never become reality. Agricultural experts, including those of the FAO, are sceptical about cultivating certain crops, like cereals, domestically, and regard an import-heavy strategy as more rational for tiny and rich Qatar.[61] Technology choices and the eventual scale of application, as well as economic feasibility, seem to bear significant uncertainties—not forgetting the environmental impact of a massive increase in desalination in the shallow Gulf waters. While an impressive blueprint for sustainable agriculture and food security for a small, arid and wealthy Gulf country indeed looks likely to materialize by the mid-2010s, it looks uncertain whether this could become a model replicable anywhere else, as the programme envisages. Also, many other factors, already outlined elsewhere in this book, threaten the successful implementation of QNFSP's ambitious plans, including Qatar's natural gas abundance and existing rentier structures; the contingencies associated with highly concentrated decision-making power; the challenges of inter-institutional coordination and cooperation; and the GCC states' weak track record in timely implementation of visionary plans. What is, however, certain is that the QNFSP is already preparing the ground for a strong, overarching national response to some of the major natural unsustainability challenges Qatar will face in the coming decades, brought about by the complex interactions of a hostile climate, unsustainable resource consumption, fossil fuel dependence, and, ultimately, climate change.

Contribution to climate change and mitigation actions and policies

Especially since the late 2000s, Qatar's high per capita emissions have been the source of much unwanted international attention and the impetus for defensive policy positions in the UNFCCC. Qatar's per capita CO_2 emissions, as estimated by the World Resources Institute (44.6 tonnes in 2008), are indeed the world's highest: roughly 60% higher than Kuwait's, ranking second. The country's total and historical contributions (0.21% and 0.11% in 2008), like those of the smaller GCC states, are minor on a global scale. Nevertheless, total emissions, at 62.2 Mt of CO_2 in 2008, have been growing fast as a consequence of the expansion of the hydrocarbon industry and economic and population growth, with the associated housing and energy requirements. After producing extremely modest emissions until the 1980s, between 1990 and 2008 Qatar had among the world's fastest greenhouse gas emission growth rates, on average 8.4% per year, while the MENA and GCC averages were 4.7% and 5.3% per year, respectively.[62]

Qatar's first emissions inventory was finished in 2009, but only published two years later. Unpublished preparatory data which looked at the years 2001 and 2006 reveal a significant increase in emissions in this period (147%) and confirm that Qatar's emissions originate from the hydrocarbons industry. The published emissions inventory estimates the country's total greenhouse gas emissions at 62.4 Mt/CO_2e in 2007. In a very close parallel to Abu Dhabi, where ADNOC produces over half of total emissions, Qatar's oil and gas industry alone accounted for 50% (59% upstream, 11% downstream) of total greenhouse gas emissions from the energy sector in 2007.[63] In a clear attempt to highlight the share of Qatar's export industries in total emissions, the GSDP breaks up the figures differently, attributing 67% to energy and heavy industries (37% oil and gas; 12% flaring; 8% petrochemical and cement; 10% industrial electricity and water consumption) and 33% to households and commercial users (18% domestic and commercial electricity and water consumption; 7% transport; 6% construction; 2% other).[64] Despite the lack of more recent data, for 2008–11, it can certainly be asserted that both the doubling of the population from around 900,000 to 1.7 million, and the coming on stream of a number of LNG trains, have brought about a record-high growth in Qatar's total emissions in this period.[65]

The inventories also show the significance of reducing gas flaring for domestic emission cuts: despite a nearly fourfold absolute growth in emissions from oil and gas operations (not including flaring) between 2001 and 2007, both the share and the amount of flaring dropped dramatically, from 45% (18.1 Mt/CO_2e) to 12% (7.5 Mt/CO_2e).[66] In 2009, Qatar was the first GCC state to join the World Bank's Global Gas Flaring Reduction Partnership, established in 2002, with a national aim of achieving zero flaring during normal operations by 2010—which obviously still remains to be achieved.[67] In 2011, according to government data, Qatar's potential to cut emissions by optimizing energy use and cutting flaring and venting amounted to 8 Mt/year.[68] Qatar's most important mitigation project to date has been the al-Shaheen flaring gas recovery project with Maersk Oil of Denmark. The project was registered as a UN Clean Development Mechanism (CDM) project in 2007, and its annual emission reductions of 2.5 Mt/CO_2e/year would have been worth US$41m at 2009 prices,[69] had Qatar decided not to issue the credits—because, it was said, the emirate did not wish to appear to be taking advantage of the UNFCCC, with which it has had a complicated relationship. In 2010, the Ras Laffan Industrial City initiated a US$1bn project to minimize gas flaring at LNG berths.[70]

In the latter part of the 2000s, a number of other initiatives emerged in the areas of clean tech and renewables R&D and investments, which came to characterize climate change-related efforts in Qatar in this period. Most of these were linked to the Qatar Science and Technology Park (QSTP) under the Qatar Foundation and Sheikha Mozah, which will be discussed below. Although still at a very early stage, the Heir Apparent's food security programme is planning large-scale renewable energy-based desalination. There are reports of Qatar University studies on using solar energy in agriculture from the 1970s.[71] More recently implemented small-scale solar projects have included street lights for a waste management facility in Ras Laffan and bus shelters on Doha's Corniche.[72] There have also been some announcements of bilateral and multilateral cooperation: in 2007 Qatar, like the UAE and Kuwait, pledged US$150m to the Saudi-initiated OPEC clean tech fund. In 2008 the Qatar Investment Authority invested £150m in a £250m joint clean energy and technology fund with the British government.[73] In 2010, the Qatari Energy Ministry and US and German counterparts signed MoUs on technology cooperation in renewable and alternative energies.[74]

In the area of sustainable building, the local developers Barwa and Diar developed a green building rating system, QSAS, which they pledged to implement as of late 2009.[75] Reflecting a gas abundance-induced, rather lax attitude to broader implementation, according to the GSDP, the standards are expected to become a mandatory requirement for all government buildings by 2016, and later on extended to all *new* residential and commercial buildings.[76] Also, as of 2010, Qatar Cool has provided district cooling for forty-seven residential and commercial towers in the West Bay area of Doha, and the company will ultimately provide cooling for the entire Pearl artificial island development.[77] Qatar Foundation's Education City too incorporates district cooling. Qatar's winning bid for the 2022 FIFA World Cup includes 12 futuristic outdoor stadiums, which will have solar panels and the 'world's first' carbon neutral cooling system.[78] In the transport sector a US$29bn national rail and a 300 km metro network in Doha are also expected to be nearly completed by the time of the World Cup.[79] On the downside of the massive development projects, even if the country's new construction and megaprojects will be green or sustainable by any criteria, they will still require large amounts of energy and resources to construct. Also, as elsewhere in the GCC, retrofitting of buildings to increase energy efficiency still remains a largely uncharted area, owing to the utility subsidy regimes and abundance of lower hanging fruit, in the form of new projects, in the construction sector.

By early 2012, Qatar still had no official domestic climate policy, apart from the three aspirational aims in the National Vision 2030 and the commitment to halve the *intensity* of emissions from flaring from 23 mcm/Mt of energy produced to 11.5 mcm/Mt by 2016, pledged in the five-year development strategy.[80] The GSDP's second *Human Development Report* lists a number of actions as Qatar's responses to climate change, including the Qatar National Vision 2030; investments by the Qatar Science and Technology Park and the Qatar National Research Fund; participation in the Kyoto Clean Development Mechanism; the national committee for climate change; participation in OPEC's clean tech fund; education; and air quality and emissions monitoring actions.[81] Notably, the UNFCCC communication from 2011 stresses that Qatar will pursue *voluntary* emission reductions only 'as long as they are in line with sustainable development'.[82] Local authors have argued that Qatar's CO_2 emissions could be significantly reduced without compromising economic growth, especially in the oil and gas, energy use and transport sectors, through energy conservation, efficiency and 'adjust[ing] the value structure of the society'. Still, the success of existing plans and aspirations, as noted by local stakeholders, will also depend on inter-institutional coordination and cooperation[83]—and not the least, consistent implementation.

Whereas since around 2007 the leadership in Abu Dhabi, and also to some extent in Dubai, has felt an urgent need to address the high domestic natural resource consumption and greenhouse gas emissions—for energy security, environmental integrity and image reasons—the top elite in Qatar have not quite shared this sentiment. Firstly, there has been no high-level elite member, with a clear environment-related vision, available to take up this niche of patronage. Moreover, neither have there been other important domestic actors with a vested interest in a 'greener' image (such as those involved in Abu Dhabi's Masdar), nor has there emerged a sufficient number of domestic projects in the area of alternative energies and technologies to transform alternative energies, environmental sustainability or climate change mitigation into an area of national interest. Consequently, until November 2011, when Qatar won its bid to host the 2012 UNFCCC Conference of the Parties, there was no evident pressure to draft a domestic climate change strategy. Owing to Qatar's robust domestic energy security situation, there was no urgent need for near-term alternative energy plans or targets either.

In general, natural gas has been the 'perfect excuse' for Qatar: not only does it provide domestic energy security, it has also allowed the elite to evade accusations regarding the high per capita greenhouse gas emissions. The fact that natural gas is the cleanest fossil fuel, and that two thirds of Qatar's emissions come from the export industry, formed the core of Qatar's external climate policy, led until 2011 primarily by the Ministry of Energy and Industry, as will be discussed in chapter 8. Given the changes in the Energy Minister portfolio the future leadership of Qatar's climate governance was still wide open as 2011 ended. While until early 2011 Energy Minister al-Attiyah had been active in promoting natural gas and to some extent clean fossil fuel technologies, the new Minister al-Sada, more of a technocrat, has still to build a profile for himself, although close observers suggest he is more open to exploring renewables than his predecessor.[84] Towards the late 2010s, Sheikh Tamim and his food security programme might well emerge as the domestic pioneer in renewables. If all else fails, Sheikha Mozah's Qatar Foundation, through its R&D-driven approach, might under optimal conditions spur a gradual but in many ways more sustainable response to the challenges of the newly emerging global energy paradigm.

The case of Qatar Foundation and the QSTP

Owing to its centrality in Qatar's domestic climate change-response of the late 2000s, the Qatar Foundation, and particularly its Qatar Science and Technology Park (QSTP), form an interesting case study and point of comparison with Abu Dhabi. Despite their frailness, the Park and its mother institution, the Qatar Foundation, generated the most advanced early efforts at enhancing Qatar's capabilities and actions in the areas of alternative energy, environmental sustainability and climate change. Coming directly under the patronage of Sheikha Mozah, these projects heavily focus on building domestic expertise and technology know-how in a longer term, rather than bringing immediate changes. Taking the slow road might turn out to be more sustainable than buying existing technology and models of implementation. The big question, however, remains whether this is too slow for the prevailing rentier mentality.

Established in 1995, after Emir Sheikh Hamad bin Khalifa's coup, the Qatar Foundation for Education, Science and Community Development is officially described as 'a vehicle to convert the country's current, but temporary, mineral wealth into durable human capital'.[85] Kamrava describes the Foundation as 'by far the most comprehensive and ambitious of… government-controlled NGOs', through which state-society links are strengthened and the Emir's power is consolidated at the same time. He has suggested that such institutions also function, aided by their 'deliberately vague' status, as 'penetrative arms of the state [in] typically potential centres for the formulation of anti-state anger',[86] such as academia.

Sheikha Mozah bint Nasser has been Qatar Foundation's chairperson since its establishment, granting the Foundation 'seemingly limitless' resources under her patronage.[87] In late 2011, in a reshuffle of the orga-nization's board structure, she took an important step to consolidate her role and strengthen the Foundation's continuity by establishing a new board of trustees, almost fully composed of her sons and daughters (four in total), naming Tamim (the Heir Apparent) as its vice chairperson, and appointing her daughter Hind as the vice chairperson of the board of directors. The long-time vice chairperson, Saif Ali al-Hajari, also the chairman of the Friends of the Environment Centre, seems to have been quietly sidelined. By 2010, the enterprise had mushroomed to over forty member organizations and concepts in the areas of education, science and research, 'community development' (culture, heritage and social ser-vices), and related business joint ventures. These included institutions as varied as a riding academy, a children's channel, a solar technology ven-ture, a national research fund, and a faculty of Islamic studies.[88] The Foun-dation has also brought to Qatar branch faculties of well-known American universities that offer scholarship-sponsored degree programmes for both Qataris and non-Qataris. These have since 2011 operated under the patronage-exuding title of Hamad bin Khalifa University. Facilities for the universities are free of charge and operational costs are fully covered by the Foundation.[89] Most of the Foundation's member institutions are located in Education City, inaugurated in 2003, which comprises a 14 km^2 campus on the outskirts of Doha, dotted with architectural master-pieces by world-famous architects. Reflecting the enterprise's vague sta-tus, QF does not publicize its budget, nor has it ever produced a financial statement. Some sources have suggested that Education City has a bud-get of US$6.6bn, but the massive Sidra Medical and Research Center built in the area is reported to cost US$7.9bn alone.[90]

Knowledge-economy building is at the core of the Qatar Foundation's activities. In addition to graduate programmes and research undertaken in the hosted universities, a regional branch of the American RAND Corporation has established itself under the Foundation, and has been responsible for a controversial K-12 educational reform. QF also man-ages the multi-million Qatar National Research Fund, aimed at research proposals in areas defined as of national interest. The QF Research Divi-sion, set up in 2007, has also set up three independent research centres in the fields of health, computing, and energy and the environment, with an ultimate vision of establishing a robust scientific research network in Qatar. During its first few years (2012–16), the Qatar Environment and

Energy Research Institute (QEERI), the most important of the three, will focus on four areas of priority: sustainable building, ecology, agricultural research and material and nanotechnology research, all well in line with emerging domestic priorities, including solar desalination and water and food security. Applied and policy research into alternative energies/technologies is also carried out in other QF-affiliated institutions, most prominently Texas A&M, as well as outside QF, most notably by the Environmental Studies Center of Qatar University.

The Qatar Science and Technology Park: energy and sustainability developments

Technology development is another key area of the Foundation. The Qatar Science and Technology Park (QSTP), the Education City's R&D-oriented business incubator, was initiated in 2002 by Sheikha Mozah and inaugurated in 2009.[91] With expected synergies from the other QF institutions, it aims to advance the commercialization of the Education City's innovations, create jobs and build up local knowledge industries by seeking to attract foreign research institutions and companies to establish a presence or a joint venture in the Park.[92] It shares the same research foci as the three QF research institutes, with most funding received by the Park's two energy clusters: hydrocarbons and alternative energy.[93]

The QSTP is a free zone that provides companies with facilities, services, and networking and commercialization support, allowing for 100% foreign ownership of companies, for a period of 1–20 years.[94] By 2009, it had received US$600m in funding commitments from the Qatar Foundation and over US$200m from partner companies, including EADS, ExxonMobil, GE, Shell and Total. In early 2012, the QSTP listed forty-eight member companies.[95] The QSTP's executive chairman, Tidu Maini, who also serves as Sheikha Mozah's technology and science adviser, is an Indian-born businessman and academic who previously served as Imperial College London's head of technology transfer.[96] Through its members, the QSTP has quickly grown into Qatar's (still small) centre of gravity in the areas of technology development of carbon capture and storage (CCS) technologies, energy efficiency, alternative energy and technologies, and sustainable building standards. In the area of solar energy the Park is already involved in research at both ends of the value chain: through its polysilicon project and through its solar technologies testing programme.

Research related to carbon capture and storage evolves through a 10-year joint research collaboration between the QSTP, Imperial College, Qatar Petroleum and Shell (funded by the two latter) since 2008 to study local carbonate reservoirs and CCS technologies, and to build domestic engineering capacity. The US$70m project is carried out by Qatar Carbonates and Carbon Storage Research Centre at Imperial College London, which has recruited new academic staff, including forty PhD students and post-doctoral researchers, some of whom are Qataris, for the project.[97] It was still at its early stages in early 2012, when no date for carbon injection demonstration projects had been announced. Nevertheless, given the massive potential of CCS in GTL and LNG processes and enhanced oil and gas recovery,[98] this area will doubtless receive increasing attention from the government, as carbon taxes, quotas and trading become globally more common.

Energy efficiency expertise is developed under Chevron's Centre for Sustainable Energy Efficiency, established in 2011. Equipped with an initial five-year budget of US$20m, and hosting a visitor centre, the centre also works with a local company on a solar test site.[99] The experimental solar facility for the study of solar-to-electricity conversion methods was first announced at the Park's inauguration in March 2009 by Green-Gulf, a Qatari-Saudi advisory business led by a young Qatari, Omran al-Kuwari. Chevron joined the Qatari-initiated project in 2010 when both sides agreed to invest up to US$10m in the project over a period of up to four years. The construction of the 35,000m^2 test site finally began in late 2011. In its first phase, the project will test the performance of different solar technologies in Qatar with total systems installed at approximately 500 kW. This will eventually be fed to some of the QSTP's buildings.[100] GreenGulf has also developed jointly with Abu Dhabi-based Environmena a 700 kW rooftop PV system for the new Qatar National Convention Centre, and installed solar panels in four Qatari schools.[101]

At the other end of the value chain, Qatar Solar Technologies, a joint venture of QF (70%) and the German company SolarWorld (29%), established in 2010, will build a facility that converts natural gas into polysilicon, used as material for solar panels. With an initial investment valued at US$500m, the venture is clearly the largest investment in renewable energy technologies in Qatar's history. The first part will see the construction of a polysilicon plant in Ras Laffan, expected to be completed by 2013 and to create 300 new jobs. The plant, the first in its kind in the

Middle East, will have a large capacity for a start-up company, 8,000 tonnes/year. The company hopes to benefit from the strong and rising international demand for photovoltaics, domestic availability of cheap electricity and Qatar's location between the growing markets in Asia and Europe. In the absence of domestic demand, however, the factory's output will be sold abroad, mainly to Germany.[102]

Smaller-scale alternative energy projects within the QSTP framework have included a feasibility study for producing aviation fuel from algae. It has also assisted two academic institutions in research on the solar-cracking of methane for hydrogen generation.[103] Sustainable building is also covered by the Park's activities. By 2010, there were two QSTP members working in the sector: TCE Consulting Engineers and the Barwa and Qatari Diar Research Institute. The former, part of the Indian Tata Group, announced in 2008 that it would invest US$12m in developing integrated software for sustainable buildings and a blueprint for a solar thermal power station over the next five years.[104]

The Barwa and Qatari Diar Research Institute (BQDRI), later renamed as Gulf Organisation for Research and Development (GORD), was established in 2009 as part of Barwa Knowledge, the corporate social responsibility platform of the Barwa Real Estate Company, together with the Qatar Investment Authority-owned Qatari Diar Real Estate Investment Company. The institute, led by a former Barwa executive, Yousef al-Horr, engages in research and education and training for the construction industry, and has developed a performance-based sustainability rating system devised for Qatar's climatic and other conditions. Originally launched in 2009 by Barwa and Diar, together with a centre of the University of Pennsylvania, the Qatar Sustainable Assessment System (QSAS) prioritizes energy and water consumption, as well as cultural and economic value considerations, and rewards local building materials and architecture inspired by Qatar's heritage. In addition to administering the QSAS, GORD engages in a range of sustainable building-related research topics and consulting and training services, and also plans to work on a Regional Sustainability Assessment System for the GCC and the Middle East. As an innovative approach, GORD has planned incentives for projects in Lusail City in the form of increased land allocations in relation to the number of stars they receive.[105]

Unlike Abu Dhabi's Estidama, which is an adaptation of the American LEED system, QSAS is, according to its authors, a result of extensive international benchmarking and has been relatively well received not only by the two founding developers, but also by other Qatari institutions. In 2009, Barwa and Qatari Diar announced that they would apply QSAS to all new projects, Lusail City, as the first one.[106] The Barwa City development of 128 apartment buildings in Doha will also receive a QSAS rating. In 2010, the Public Works Authority Ashghal made a similar announcement regarding its public building projects, and GORD signed an MoU with Kahramaa on 'the provision of measures to create a sustainable built environment'. Kahramaa is also building a QSAS-awarded water and electricity Awareness Park in Doha, expected to open in 2013.[107] Perhaps most importantly, the system was incorporated in mid-2012 in Qatar's building code, with an eye on making it compulsory for all new public buildings starting from 2013. Also, seeking to expand to new dimensions, the name of the rating system was changed to GSAS (with 'g' for Global), which GORD proclaims to be 'the standard for excellence on sustainability in the MENA region'. Boding well for the concept, Kuwait has already shown interest in adopting the system.[108]

Other sustainability-focused QSTP member companies include four major energy giants, ConocoPhillips, ExxonMobil, Shell and Qatar Petroleum, each of which has entered the QSTP with a medium-size research institute, heavily suggesting their much-speculated function as offset projects. ConocoPhillips, together with GE, has established its Global Water Sustainability Centre that concentrates primarily on industrial water sustainability, particularly methods for treatment of water from oil production and refining, and includes a visitor centre aimed at youth awareness raising.[109] ExxonMobil Research Center Qatar received in 2010 a five-year investment pledge of US$60m. The centre concentrates on LNG technologies, sustainable industrial water management, and carbonate reservoirs. In 2008 Shell opened its Research and Technology Centre, and it has announced spending plans of up to US$100m through to 2018. Projects include two joint endeavours on synthetic jet fuel for aviation with a consortium of partners. Qatar Petroleum's Research & Technology Centre, with a budget of US$75m over its first five years, works mainly in the area of upstream oil and gas operations, but the environmental impact of industries and climate change, and energy efficiency, are also on the centre's agenda.[110]

Other developments under the Qatar Foundation umbrella include a university-based research institute, research partnerships with domestic and foreign institutions, and a number of construction projects seeking sustainability certification under the competing American LEED rating system. Upon arrival to Education City in 2003, Texas A&M University established a Sustainable Energy Research Laboratory, which has been studying, among other things, the use of solar power for the production of hydrogen.[111] Sheikh Tamim's food security programme QNFSP is also conducting research on solar and water technologies with Qatar Foundation, QSTP, Qatar University, and Texas A&M.[112] Partnerships with regional peer institutions, like King Abdullah University of Science and Technology in Saudi Arabia, have been explored and encouraged by Qatar's RAND branch, which has been advising QF in strategic development.[113] QF is also seeking to educate the students, through a number of smaller sustainability campaigns and projects taking place at Education City. Education City is starting a recycling scheme for all its buildings, but implementation is taking longer than expected, since at the request of the Ministry of Municipality and Urban Planning, QF has decided to wait for a new solid waste management centre near Mesaieed to begin receiving its waste. Previously, Qatar has only had one private facility for paper and carton recycling and one for plastics, and a full-scale collection scheme for recyclable domestic waste is yet to materialize. At some point, recyclable waste was reported as being accumulated in air-conditioned storehouses in Education City—an indication of the difficulties of introducing a new lifestyle paradigm.[114] In the other end of the scale, headway was made in state-level cooperation and tech transfer when the QSTP and the US Department of Energy signed an MoU in 2010 on co-developing projects in the areas of energy efficiency, CCS and solar technologies.[115]

Sustainable building developments with links to the Qatar Foundation include a number of buildings at Education City and the QF subsidiary Msheireb's 'revitalization' project. It is reported that the Foundation's board decided as early as 2004 to seek LEED certification for all new buildings at Education City.[116] A few years later, some steps in this direction were taken: a residence hall complex with capacity for 600 students, expected to be finished in 2012–13, is seeking to become the world's first platinum LEED certified project in its category, with features such as zero waste, motion-sensitive lighting, solar panels and grey water filtration.[117] Also, parts of the Qatar National Convention Centre in Education City, designed by the world-famous Japanese architect Arata Isozaki, have been built for LEED gold certification. The rooftop solar panels of the US$1.2bn centre, opened in 2011, produce 12.5% of the building's energy requirements.[118] The Msheireb project, announced in 2009 and to be completed in 2016, with a price tag of US$5.5bn and an extension of 35 hectares with 226 buildings, will also seek LEED gold certification. The project includes an underground section, with pedestrian areas planned to reduce car use.[119] Other QF LEED projects will include educational facilities in the Doha area and neighbouring towns.[120] In addition, with the goal of making Education City car-free by 2015, a US$100m eco-efficient tram project, comprising 11.5 km of tramways, was awarded to Siemens in mid-2012.[121]

The QSTP's functions

While it is still too early to judge the QSTP's successes and failures and whether it will leave a mark on Qatar's future knowledge economy, it has certainly avoided the rollercoaster ride of Masdar by not promising too much too soon. On the contrary, owing to the relatively narrow scope of their mandate, the QSTP and other research activities under the Qatar Foundation have so far remained in the shadow of Education City's prestige universities, Qatar's hyperactive foreign policy, and other areas of external branding. Official press material and speeches spell out the Qatar Science and Technology Park's role and function in the Qatar Foundation's grand strategy and, more widely in the Qatari economy, its role as an engine for accelerating QF research, which focuses on specific technologies and areas of national priority.[122] The Park declares that its purpose is to help build Qatar's post-carbon economy by becoming a 'recognised international hub for research, innovation and entrepreneurship'.[123]

Despite the obvious differences between the QSTP and Abu Dhabi's Masdar Initiative, these two make an interesting comparison. Differences are evident at least in the level of funding, breadth of activities, scope of investments, speed of implementation and importance given to R&D. While the QSTP focuses on a number of technology sectors, instead of Masdar's one, the Park's scale and dimensions are clearly smaller, with 'only' US$600m committed in the first phase by QF. Masdar had by mid-2009 already invested US$3bn in alternative energy and sustainability projects both domestically and abroad.[124] The QSTP's focus is heavily on small-scale, bottom-up technology and knowledge transfer: projects concentrate on 'technology development'[125] and these are undertaken by institutions and other joint ventures that have a presence on the Park's premises. Masdar is aiming at working with a more diverse combination of subsectors, technologies and projects, including clean tech funds and foreign investments, a domestic PV industry and different types of solar energy plants and installations, CCS and CDM projects, hydrogen energy applications, real estate and research. While Qatar's strategic decision has been to avoid moving quickly into pilot projects and large-scale implementation, Abu Dhabi's Masdar initially lost no time, although it has been forced to slow down significantly since then. QSTP members' pilot projects, like GreenGulf's 500kW solar testing site and the Shell carbonate reservoirs project, are still relatively smaller and at very early stages compared with those in Abu Dhabi, such as the solar plants and the CCS pilot projects. Moreover, as a final differentiator, owing to their distinct business models and foci the QSTP and Masdar do not directly compete with each other at the level of technologies or business niches.[126] This is important, as arguably Qatar, like other GCC states, has become careful not to be seen as imitating innovative areas of branding by neighbouring monarchies too closely.

The importance that QF gives to R&D and 'domestic' innovation is the clearest difference from Abu Dhabi's Masdar. According to Qatar-based stakeholders' views, the 'content of the work' is much more important for QSTP than for Masdar; while the former has taken the best, the latter 'accepts anyone'.[127] And while the QSTP has managed to attract a number of medium-size members, despite enormous media attention Masdar has struggled to sign deals with tenants. On a more detailed level, the Park has so far clearly had a heavier focus on alternative energy R&D than on its commercialization. This, according to the company representatives (in the case of solar) indicated a premeditated strategy 'not to rush in', ordered by the former Energy Minister al-Attiyah.[128] Another aspect of QSTP's approach is its emphasis on entrepreneurship and Qatarization,[129] which are also becoming visible to some extent in Masdar's research and recruiting.

In a broader social context, the Park's investments in alternative energy and technologies are expected to bring multiple additional benefits, ranging from domestic energy security to energy efficiency and climate change mitigation, as is often assured by the Park's executives and members.[130] In general, however, environmental considerations, including climate change, clearly have a narrower role in the official discourse of QSTP's alternative energy and environmental projects.[131] On a strategic level, a number of QSTP-related energy and sustainability projects are described as responses to the National Vision 2030 document.[132] Establishing patronage is an important element in the discourse and practices of Qatar Foundation and its member institutions, as is the case for Masdar. Project launches and award-receiving speeches by QSTP's member companies often give credit to Sheikha Mozah and Emir Hamad.[133] In practice, there has been little questioning of who runs the show, with the Qatar Foundation closely identified as Sheikha Mozah's project, and senior executives have remained remarkably constant throughout the years, with the partial exception of Mozah's children joining the board structures in late 2011. Another form of legitimacy-building is the justification of contemporary sustainability programmes with a neotraditional environmentalist discourse—identical to that of the UAE's Sheikh Zayed—that imitates an imagined pre-modern, Bedouin-inspired environmentalism. In 2010, in a speech renaming the Heart of Doha project as Msheireb, Sheikha Mozah said: 'Our past clearly reflected that communities in Qatar have always been close-knit. People lived and worked together in harmony with the climate, with the land and with each other. We had our own ways of dealing with our environment which was sustainable and human in scale, and thus, our architecture reflected the unity of our family's identity'.[134] By 2012, a new text had appeared on Sheikha Mozah's personal website, representing a rhetorical alignment of the evolving legitimacy-building process with the fast-expanding environmental activities under her patronage: 'From the outset of Qatar's economic transformation, His Highness the Emir and Sheikha Moza realised that this rapid development must be environmentally sustainable. […] In so doing, Qatar also can be a leader in the world's environmental revolution by developing the green technologies to build a more sustainable future'.[135]

Future prospects

While as of 2010 the Qatar Foundation and its technology park had clearly taken the vanguard position in climate change-related sustainability efforts in Qatar, they were still isolated examples, somewhat reminiscent of Hertog's 'islands of efficiency'.[136] Their economy-wide impacts can therefore be expected to stay limited as long as climate change and sustainability remain without attention from high-level ruling family members and on the margins of the government's agenda. For the fore-seeable future, QF's and the Park's role and significance for economic diversification and domestic energy security will arguably remain marginal. Important contributions can be expected in the medium term from the member institutions' work in energy efficiency and CCS-related knowledge and possibly, in the longer term, in solar technologies. Also, if successful in developing cutting-edge energy and sustainability technologies, QSTP-based companies and Qatar could in the future achieve a regional market position in this economic niche. While Masdar has undoubtedly caught the world's attention and inspired the local elite to ride its success to a further consolidation of this specific sector of the economy, the QSTP is still just a small piece in the big puzzle of Sheikh Hamad's National Vision. And Qatar Foundation's road to a low-carbon knowledge economy is set to be long and bumpy.

The idea seems to be that Qatar has found itself choosing a different path to natural sustainability: as a manager of one of the QSTP's member institutes noted in reference to Qatar and Abu Dhabi: 'I am sure you know the story of the tortoise and the hare'.[137] Fundamentally, however, Qatar's approach to alternative energy and sustainability-related technologies shares the same fundamental paradox as all attempts at turning Qatar into a knowledge economy; expectations are high, but results will take a long time to become visible. Here the persisting rentier mentality, manifested through expectations of fast results with little effort, and the small total number of Qataris are the worst enemies. As a technology park, QSTP's continuity will ultimately depend on its success in keeping the member companies after current investing commitments (presently ten years at most) and lease terms end. Financial resources will not be the issue, but other factors might work against the ambitious plans, such as slow state bureaucracy and decision-making processes,[138] the above-described cocktail of strong and weak institutions, lack of appropriate regulation and subsidies for renewables, and in general the overriding importance and abundance of fossil fuels, in particular natural gas, in Qatar's economy. Through the QSTP and Qatar Foundation, the government is trying to build alternative energy and sustainability expertise the hard way. If it succeeds, the tortoise will indeed have beaten the hare.

8

THE GCC STATES IN THE INTERNATIONAL
CLIMATE REGIME

Throughout the two-decade history of the international climate regime, under the United Nations Framework Convention on Climate Change (UNFCCC), the OPEC group, led by Saudi Arabia, has opposed mitigation measures that would potentially harm oil exporting countries' external revenues. Abu Dhabi/the UAE and Qatar, both members of the OPEC group, have traditionally been regarded, together with Kuwait, as merely supporters of the Saudi line. But as the issue has been picked up by the foreign ministries of the former two states, a need for a more clearly defined national negotiating position has emerged, beginning from 2009 and 2012 respectively. After presenting the dynamics of the most relevant reference groups for the GCC and the role and positions of Saudi Arabia, the dominant actor in the group in the period from 1995 to 2011, this chapter examines in detail the external climate policies of the United Arab Emirates and Qatar. Most important, it demonstrates how, in the case of the UAE, changes at the domestic level interacted with the international level, creating new foreign policy priorities. This led, in 2009–10, to the emergence of a new policy leadership that reformulated the country's alignment and priorities in the climate regime. Simultaneously, in Qatar, owing to the strong ownership of external climate policy by the country's energy sector, in the absence of major domestic developments in the area the country's external climate policy remained static. It was only when the Foreign Ministry picked up the topic in its assemblage of areas of external prestige-seeking that Qatar was pushed to start devising an independent position.

Reference group dynamics[1]

All Gulf monarchies have acceded or ratified both the Convention and the Kyoto Protocol. The six joined the UNFCCC in 1994–96, and the Protocol that followed shortly entered into force in 2005–06.[2] Under this framework, also referred to as the international climate regime, their main reference groups have been OPEC, the GCC, the G77+China group, OAPEC (the Organization of Arab Petroleum Exporting Countries) and, to some extent, the Arab League. The locus of policy coordination and position formation from the small Gulf states' point of view lies within the first three of these groups. Because of its opacity, little is known by outsiders of GCC states' coordination in this area. The group has generally been portrayed as 'well-disciplined, with a unified policy', its smaller member states echoing Saudi statements.[3] Stakeholders and personal observations however suggest that this is only partly true, and quickly changing. According to a delegate of Bahrain, that country supports Saudi Arabia in all decisions;[4] indeed, foreign policy dependencies are an important factor in the case of Bahrain.[5] Kuwait has also been a near-guaranteed supporter of Saudi Arabia, but because of its OPEC membership and low domestic interest in the issue, it can be argued that it supports Saudi Arabia for reasons of interest alignment rather than conformity. Although they have often supported Saudi statements, Qatar and UAE negotiators have assured that their countries do not *follow* the Saudi position and that Saudi Arabia has never imposed its will on the smaller states. On a few occasions, the UAE and Qatar have indeed deviated from the Saudi position. Still, both have admitted often providing silent support to positions or statements presented by Saudi Arabia, particularly until 2011. This presumably happens either when the countries' perceived interests are aligned or when deviating would entail a higher political cost. In the case of the UAE, conformity has been described as having reached the point where Saudi Arabia would feel comfortable with signing on behalf of the UAE in support of a position.[6] Other differences exist too: Oman for example, unlike the four GCC OPEC states, did not support the inclusion of carbon capture and storage in the Kyoto Clean Development Mechanism (CDM).[7] There are a number of instances in which small GCC states have supported Saudi Arabia even when this has been against their interest in one way or another. A clear example is the UAE's support in the late 2000s for Saudi/OPEC positions, perceived by other parties as obstructionist, while Abu Dhabi simultaneously sought international credibility for its alternative energy leadership aspirations.

The OPEC group, although more heterogeneous in terms of its members' interests, has also been an important reference group for the Gulf monarchies as it aggregates its member states' common economic concerns, most important being those relating to potential negative impacts of international mitigation policies and measures (or 'response measures').[8] Notably, Oman and Bahrain are not members of this group, while non-member Dubai has representation through the UAE delegation. OPEC as an organization does not negotiate but provides logistical support for its member states.[9] The generally tight discipline of the OPEC group, often described as a Saudi vehicle of obstructionism, gives all its members added clout. The group has been described as a key player, even securing leadership in the G77+China from the late 1990s.[10] Historically, within the diverse group of developing countries, OPEC states and the Alliance of Small Island States (AOSIS) have frequently been at loggerheads due to their often highly divergent views regarding the aim and purpose of the Convention.[11] The G77+China is a group to which all six states belong and which represents the developing world's voice vis-à-vis the developed world. Despite their high income status the GCC states are classified as developing countries in the Convention and are vehemently unwilling to give up this status. Both in the G77+ China group and in the negotiations in general, the six monarchies' participation was for long primarily framed by OPEC coordination and Saudi leadership. The OAPEC group's policy positions too have been described by some observers as mostly formulated by Saudi Arabia and Kuwait, and also as dominating the Arab states' common position.[12] The Arab League member states' coordination is generally weak, owing to the divergent interests and strong coordination among the oil-exporting states, but has found new dynamism from a joint coordination agreement between Egypt and Saudi Arabia since around 2010.[13] As of 2010, Arab states' ministers had issued three joint declarations on climate change, all of which still were lacking noticeable implementation: the Abu Dhabi Declaration on Environment and Energy (2003), the Arab

Ministerial Declaration on Climate Change (2007), and the Statement on Climate Change issued by the Council of Arab Ministers Responsible for the Environment (2009). The influence of oil exporting countries on the content of the declarations is noticeable, although possibly slightly declining.[14]

Like most developing countries, Arab states, including the five smaller GCC states, have until recent years sent only small delegations to the negotiations.[15] The resource-poorer Arab states in particular, because of low domestic political prioritization, lack of human resources and the wide plethora of issues on the negotiating agenda, have found it difficult to engage in the negotiations. Even UAE negotiators have noted the same problem in relation to the large number of parallel meetings.[16] The consequent unpreparedness has led to vague positions and vulnerability to outside influence. Throughout the past two decades, Saudi Arabia has been the only Arab state attending the UNFCCC meetings with well-prepared, vocal and large delegations, which has increased the importance of its position for the other Arab countries participating with a less defined agenda and a weaker mandate. Saudi Arabia has a lot of permanent legal and technical expertise and its negotiators have been characterized as skilful and very strategic, especially on the issue of response measures. Besides their explicit or implicit support for Saudi or OPEC/OAPEC states' positions, other Arab states have generally been engaged in only some agenda items: in 2008, Algeria was described as active when it came to carbon capture and storage (CCS) and Egypt as pro-adaptation because it fears submergence of the Nile delta due to rising sea-levels.[17]

The participation of the smaller Gulf monarchies in the coordination of policies and positions under the UNFCCC framework has had different forms and intensities. In 2008, a long-term observer described them as generally 'invisible and quiet', and Kuwait and Qatar particularly as using the same rhetoric as the Saudis.[18] The most active ones in the plenary meetings have been the three OPEC member states among the GCC: Kuwait (especially in the late 1990s), Qatar and the UAE.[19] In older literature, Kuwait is often mentioned as the second leading state in the OPEC core negotiation group, with identical positions to those of Saudi Arabia.[20] The two non-OPEC states, Bahrain and Oman, have been even less active, with extremely small delegations and generally standing in line with GCC OPEC states' proposals and positions when expressing their views.[21]

Saudi Arabia and the OPEC group[22]

Fortunately, because of its candid statements and style, Saudi Arabia's interests, demands, strategy and tactics in the negotiations have been relatively well studied. Notably, however, the Saudi position has often been hard to distinguish from the OPEC position, as the latter has been extensively influenced by the former. Saudi Arabia's clout in the international climate negotiations under the UNFCCC framework is much greater than its total greenhouse gas emissions would suggest (1.26% in 2007).[23] This is due to a perception among the energy sector officials in charge of the Saudi climate policy that international action to abate climate change poses a bigger threat than climate change as such. From this it has followed that an ambitious agreement to cut CO_2 emissions has not been considered in the country's interests. Depledge has argued that Saudi Arabia's influence in the climate regime stems from a long-term strategy of obstructionism: obstructionists fear the agreement others might reach and therefore join negotiations so as to prevent it from emerging.[24] Many other observers too have convincingly demonstrated that one of Saudi Arabia's main motives in the negotiations has been to slow down the process.[25] Vihma, for example, has noticed that the country 'is specialized in provoking conflicts in the intersessional meetings [organized between the annual conferences of parties], while staying out of the media spotlight'.[26] The kingdom is also known to have close connections with the United States and shares parallel interests with its oil industry, both within and outside the UNFCCC framework.[27]

'Discrimination' against carbon dioxide and fossil fuels is a recurring Saudi theme that reflects the country's disapproval of any constraints on global oil consumption. To that end, throughout the early 1990s the country, together with Kuwait and the rest of OPEC, concentrated on stressing the scientific uncertainty of the anthropogenic (human) causes of climate change. In late 1996 the focus shifted to the adaptation side. Since then, calls for compensation for potential losses in oil revenue have been one of the main pillars of the Saudi/OPEC negotiating position.[28] This demand is rooted in article 4.8 of the Convention, which states that parties 'shall give full consideration' to actions necessary 'to meet the specific needs and concerns' of developing countries, including oil revenue dependent countries, 'arising from the adverse effects of climate change and/or the impact of the implementation of response measures'. Also, articles 2.3 and 3.14 of the Kyoto Protocol state that Annex I parties (developed countries) shall strive to implement policies, measures and commitments in a way that minimizes adverse social, environmental and economic impacts on developing countries. Oil exporters have interpreted these articles to mean they should be compensated for losses in oil revenue. OPEC's demands regarding the response measures issue are argued to have significantly hindered progress on the entire adaptation agenda, as the OPEC members have exploited the consensus mechanism applied in all decision-making, linking progress on the agenda to progress on response measures.[29] Saudi Arabia and other OPEC states are also well known for seeking to 'smuggle' the response measures issue into as many parts of the agenda as possible.[30]

On a more rhetorical level, studies that predict economic losses for the OPEC states and calls for economic compensation have been persistently applied as a rhetorical tool by the groups' member states throughout the history of the climate regime. For example, in 2005 Saudi Arabia implied that it should receive a lump sum payment of US$100–200bn to offset economic losses caused by Annex I (developed country) response measures over the period 2000–30.[31] Despite this, the OPEC states have come to understand the practical and political unfeasibility of their demand[32] and, behind the rhetoric, are known to instead demand technology transfer (CCS), as well as other less tangible issues, like assistance to economic diversification.

In the late 2000s, the Saudi negotiating position evolved around four main pillars. In addition to opposing all measures that would limit the global demand for and price of oil, and calling for compensation for losses in oil revenue, two new issues that emerged in 2005–07 were promoting so-called clean fossil fuel technologies, particularly CCS, as a win-win solution, and opposing any commitments or targets for developing countries and differentiation within the existing developing country group.[33] Around the mid-2000s, when carbon capture and storage rose on the negotiating agenda as a new potential tool for emission reductions, practically all OPEC members began actively promoting it both as a technology transfer item and as a new methodology under the Kyoto's CDM, so as to make R&D and pilot projects economically feasible.[34] In addition to the potentially massive removals of CO_2, the oil exporting countries wish to gain parallel benefits: offsetting part of the development costs of the technologies and applying CCS simultaneously for enhanced oil recovery.[35] Because of the need for developing countries also to curb their emissions to prevent dangerous climate change,[36] during the negotiations on the post-2012 climate treaty, a major political issue has been whether high income developing countries, including the three small GCC OPEC member states, should 'graduate' to Annex I or a similar group of countries with binding emission caps. However, the criteria for 'dividing' the G77 group were highly disputed and the developing countries were strongly against any binding caps because these were seen as 'a potential cap on their growth'.[37]

The institutional and human dimensions

A major contributor to the importance of the GCC coordination and the largely uniform positions has undoubtedly been the activeness and dominance of oil-related institutions, as well as key individuals representing them. At least until 2009, all GCC states' UNFCCC policies were in practice led by the respective energy ministries.[38] The Saudi oil sector, in particular the Ministry of Petroleum and Mineral Resources, has played the most important role. Throughout the 2000s, roughly half of all Saudi delegates came from this sector. Apart from reflecting the kind of emphasis Saudi Arabia gives to the issue of climate change, this also determined the kinds of interests represented.[39] Furthermore, there are a number of long-term negotiators in the GCC OPEC states' negotiating delegations, many of them from the energy sector, who have long, personal-level relationships with other GCC negotiators with similar backgrounds, and who, because they come from similar institutions, tend to agree with each other.[40] Arguably, the four GCC OPEC member states' oil sector representatives are also better informed and organized than their colleagues from other sectors, owing to their participation in the OPEC and OAPEC groups' coordination.

Diverging domestic interests

Despite obvious shared interests, there are a number of areas in which Saudi interests and those of the other Gulf monarchies are becoming increasingly misaligned. With the simultaneous signs of fragmentation in the G77+China group, as well as in OPEC (Venezuela and Ecuador, in particular), these are expected to become increasingly visible in coming years. First, Bahrain and Oman have substantially less remaining oil (and oil revenue) than the four GCC OPEC member states. Arguably, these two would therefore benefit more in the long term from advancement of the adaptation agenda than from compensation for response measures-related losses. They also stand to gain from increased capacity-building support, which would enable them to enhance domestic mitigation and adaptation responses and participate more actively in the regime, for example through the CDM.[41]

The three smaller GCC OPEC states are in a slightly different position, as in theory they have sufficient funds to quickly mobilize major domestic projects and plans, if they so wish. Qatar, which has a small population and is gaining an increasing proportion of its external revenue from natural gas (considered as a transitional fuel), arguably has less to be concerned about than Saudi Arabia (with mainly associated or sour gas reserves).[42] Kuwait's position will remain a question mark for some time on account of the constraints of its political system and its limited developments by the early 2010s in the area of alternative energies and technologies. As for the UAE, being associated with the climate regime's main obstructionist has arguably presented it, since the establishment of the Masdar Initiative, with an image dilemma: if Abu Dhabi wants to be seriously considered as the alternative energy leader of the Gulf, having OPEC as the primary reference group is not very helpful.

Although the material interests of the GCC OPEC states vis-à-vis the UNFCCC have so far largely related to the response measures agenda, most have begun exploring the diverse 'positive' resources available through more progressive participation. These include at least: funds and technology transfer through CDM-type projects; energy security and efficiency and economic diversification through embracing a low-carbon development trajectory; and intangible legitimacy resources and prestige offered by proactive policies, for example gaining the status of the region's climate champion.[43] Despite many countries and groups initially opposing the inclusion of CCS as a CDM methodology,[44] in 2010 the UAE and others skilfully broke the deadlock and, as a consequence, in the coming years, many Gulf monarchies also stand to gain from this new source of external rent. Also, in a broader context, since the mid-2000s the global shift towards increased multipolarity and the geopolitical openings in the post-Saddam Middle East have created spaces in which the smaller Gulf monarchies can seek leadership, independent of Saudi Arabia. Qatar's independent foreign policy and Abu Dhabi's nuclear energy programme are clear signs of this.

As will be argued below, broader changes in domestic priorities—which in these political economies penetrated by external influence and interests, almost invariably interact with global trends and tendencies—are important in initiating changes in foreign policy. Even Saudi Arabia has massive domestic alternative energy plans which, according to early announcements, will include US$100bn invested in nuclear and solar energy each by the 2030s.[45] At some point in the near future these developments will inevitably clash and lead to a rethink in the country's existing external climate policy. Early signs of this were visible with the removal of the long-term Saudi chief negotiator Mohammed al-Sabban in 2012. Elsewhere, this is already happening. As will be demonstrated below, in the case of the UAE, changes in state positioning in the UN climate talks from 2010 onwards have a strong link to domestic priorities but were also influenced by changes in the international and regional contexts and corresponding elite perceptions of the national role. Also, as will be shown further below, in the case of Qatar, a static external climate policy remained in place until 2012—despite an independent foreign policy orientation and a favourable energy demand security position—owing to the absence of a strong domestic group with interests aligned with a more progressive external climate policy.

The case of Abu Dhabi: IRENA as a domestic-level driver[46]

In 2009, Abu Dhabi successfully campaigned to host the headquarters of the International Renewable Energy Agency (IRENA). This constitutes the first example of the way perceptions among the UAE's key elite members regarding the international energy and climate-change agendas rapidly changed as a consequence of realizing the international political capital produced by Masdar. The second example, the broadening of the UAE's role perception in the UNFCCC, will be discussed further below.

In January 2009, IRENA was established in Bonn. Originally a German initiative,[47] its mission is to support and advance the use of renewable energy in both industrialized and developing countries. Its establishment was widely seen as a result of discontent with the International Energy Agency in promoting renewable energy.[48] Among the seventy-five countries that signed the establishing treaty were four OPEC member states, Algeria, Iran, Nigeria and the UAE, the last-named announcing it would compete to host the headquarters of the organization. Other candidates were Austria, Germany and Denmark.[49]

By June 2009, the membership of the agency had risen to 136, and the UAE had already secured supporting statements from numerous countries and high-level personalities, such as Ban Ki-moon, Rajenda Pachauri of the Intergovernmental Panel on Climate Change, and Amr Moussa of the Arab League.[50] The vote for hosting IRENA was supposed to take place in late June 2009 at Sharm el-Sheikh in Egypt. At the last moment the two remaining contenders, Germany and Austria, withdrew (Denmark having withdrawn a few days earlier), recognizing that the majority of votes (between 92 and 101) were already secured by the UAE. Abu Dhabi, represented in the meeting by a massive delegation of fifty-five, was declared the winner. Runner-up prizes were given to Vienna and Bonn, which were chosen to host IRENA's inter-organizational liaison office and a technology and innovation centre, respectively.[51] From the Western countries' perspective, the siting of IRENA's headquarters in Abu Dhabi was a symbolic move, as they regard the participation of developing countries in climate change mitigation as vital.[52] However, victory was secured by the votes of developing countries, which form the majority in IRENA; some of their votes, a leaked US embassy cable suggests, were bought.[53] Abu Dhabi's success can also be attributed to a campaign capitalizing on the good international publicity for Masdar and portraying the UAE as a catalyst for the introduction of renewable energy in the developing world.

Abu Dhabi's campaign

Although the candidacy for the headquarters of IRENA was made in the name of the UAE, with the Foreign Minister Sheikh Abdullah representing the country, the bid was purely that of Abu Dhabi and Masdar; the other main figure behind the campaign was the Masdar CEO, Sultan al-Jaber. The campaign consisted of tours by ministers and delegations in over 100 countries over a period of a few months. Additional pleas were made in UN meetings in New York and in a ministerial meeting of the Non-Aligned Movement in Havana. The case for the UAE was formulated through a few main arguments and coupled with substantial financial promises. The UAE was said to be no less than 'geographically, politically, economically, financially and technologically in a good position' to win the bid.[54] It was noted that the Middle East had not yet hosted an international organization. It was also argued that Abu Dhabi and Masdar would set an example that would encourage other developing countries to see the advantages of renewable energy and related technologies. Masdar City provided an attractive platform for the organization, and the initiative itself was used as proof of Abu Dhabi's commitment to the cause. The official campaign declared that Abu Dhabi's candidature signalled that 'even oil-producing and developing nations can and should participate in embracing renewable technologies'.[55] Moreover, the UAE was the first state to ratify IRENA's statute, in June 2009.[56]

Abu Dhabi's offer included plans to build the headquarters in Masdar City, in a green building that will also host Masdar's headquarters when finished some time in the 2010s. (Since then the headquarters project has been substantially affected by the City's delays and cutbacks, as discussed in chapter 5.) The emirate promised to cover all the building and operating costs of the agency and to underwrite an allowance for conference facilities and the employees' immigration fees. Financial promises totalled US$135m, of which US$70m was in cash, the rest coming from in-kind support. Annual loans of US$50m through the Abu Dhabi Fund for Development were also offered for IRENA-approved projects in developing countries during the period 2009–15. The package also included twenty scholarships for IRENA-recommended students at the Masdar Institute. The offer submitted by Germany, considered generally as the toughest competitor for Abu Dhabi, included only US$6 million for setting up the agency and US$3–4.5m for annual operating costs.[57] Ironically, Abu Dhabi's oil wealth may have been the deciding factor in its victory over Europe's leader in renewable energy, which had originally envisioned the organization.

Motivations and prospects

Above all, Abu Dhabi's IRENA campaign should be seen as an effort to raise the emirate's international profile. Whereas the existence of Masdar was undoubtedly a precondition for candidacy, the visibility and synergy gains are obvious. Turning the IRENA headquarters contest into a North-South issue undoubtedly played a key role in securing the majority of votes. From an international energy-security perspective, some argued, Abu Dhabi's victory was a sign of willingness to cooperate and engage in a dialogue with the energy-consuming countries, also indicating a concern for climate change.[58] Abu Dhabi's bid is an example of skilful interest aggregation among the developing countries, rarely seen in the past, and arguably it also reflects the country's rising foreign policy and diplomatic capability. The UAE's high-level whirlwind campaign took the competing Northern bidders Austria and Germany completely by surprise. The victory is said to have been a 'seminal experience' for the UAE and a source of pride for its Foreign Minister.[59] Moreover, Abu Dhabi, yet again, took advantage of the benevolent green-energy giant narrative that turned the very contradiction of its being one of the world's largest oil exporters into a publicity asset: a US embassy cable cites Sheikh Abdullah arguing that situating the agency's in an oil-producing country would symbolize 'international commitment to renewable energy'.[60]

Abu Dhabi's campaign was followed closely by the local press in June and July 2009. The government-owned *The National* ran a number of articles covering the campaign, as well as an opinion article by Masdar's CEO Sultan al-Jaber, who attributed Abu Dhabi's victory to a successful campaign and the UAE leader's commitment to sustainable development through the Masdar Initiative.[61] The environmental legacy of Sheikh Zayed was yet again brought up as an example of Abu Dhabi's continuous long-term commitment to environmentalism. The campaign website included a citation from Sheikh Zayed on the importance of conservation in the UAE's heritage, linking this to his achievements in wildlife conservation and Abu Dhabi's zero-gas-flaring policy and to the more recent Masdar-related developments.[62] Al-Jaber also hailed the legacy in the local press: 'The IRENA success is the natural harvest of what our late leader, Sheikh Zayed, planted'.[63]

The victory also came to mark a major watershed in the UAE's climate policy. According to a Foreign Ministry stakeholder, the IRENA campaign opened Abdullah's eyes to how important climate change had become internationally. It also brought the UAE's delegation into contact with a number of new countries, particularly in Africa and the Pacific, some of which said that they considered the UAE's support for some OPEC positions in the UNFCCC problematic. In early 2010 the UAE associated itself with the controversial Copenhagen Accord, which Saudi Arabia rejected. The move was meant as a high-level diplomatic sign that the UAE intended to take an independent role in the climate regime. The sensitivity of the issue was reflected in the low-key manner in which the association took place, with no public statements or press releases. The IRENA victory is also said to have been one of the main reasons behind the establishment of the Directorate of Energy and Climate Change under the Foreign Ministry; it was then seen that the country needed a common, solid stand towards the UNFCCC—in other words, Abu Dhabi's green leadership realized they had to 'talk the talk and walk the walk'.[64] There are strong indications that the IRENA campaign was also behind the UAE's participation in the prestigious US-led Major Economies Forum initiative.[65]

Certainly, the siting of IRENA's headquarters in Masdar City will also increase the chances of Masdar's longer-term survival, as well as those of alternative energy and mitigation projects and policies in general, by adding pressure on the emirate to deliver on its promises. The presence of IRENA will also raise the prominence of those in the ruling elite who are pushing for renewables and more sustainable energy policies as a complementary source for oil-based growth. By 2012, of the four other GCC states that had signed the IRENA treaty, two had joined the agency—a small victory in terms of regional prestige. Only Saudi Arabia had not yet made up its mind.

The first years of the agency's existence in Abu Dhabi, however, demonstrated that there are at least three potential problems for the UAE's international credibility as an impartial host. First, IRENA is a universal organization, with diplomatic representation eventually to be established in connection with the headquarters; the expected presence of Israel will both constitute a major headache and a litmus test of pragmatism for Abu Dhabi's elite,[66] although there have been private high-level assurances to the US that all members will receive equal treatment.[67] Secondly, the strong role of Masdar and the Foreign Ministry, which have been providing the agency with logistical and other support, and that of the Abu Dhabi's government, which in 2011 paid nearly half of IRENA's budget,[68] will easily create the impression of a relationship of dependency between the agency and its host. Thirdly, and in relation to previous discussion, as it has an authoritarian regime where statements going against the official state line in certain issues are not tolerated, the UAE will need to ensure that the inner workings of IRENA will be respected, even if they go against the UAE's own policy. A telling sign of both the difficulties of maintaining an impartial image and the murky deals involved in the bidding campaign was the early resignation of the agency's interim director general in October 2010, which came amidst an eclectic mix of rumours, ranging from top-level mismanagement and incompetence to external and internal political pressure to resign.[69]

The case of the UAE in the UNFCCC: shifting role perceptions

In the case of Abu Dhabi, the growing conflict between the passive, Saudi-conforming climate 'policy' on the one hand, and the domestic developments and the external pressures ensuing from its hosting of the IRENA headquarters on the other, came to be seen in 2009 by high-level elite members as a potential external image issue for the state. This led to a fast, albeit careful realignment in the UAE's external climate policy, starting from 2010.

The UAE ratified the UNFCCC in 1995 and acceded to the Kyoto Protocol a decade later, in 2005. The small relative size of the country's economy and total emissions make it a small player in the climate regime, and for a long time its only tangible contribution to the negotiations came through supporting the Saudi/OPEC position. Until 2010, when the Ministry of Foreign Affairs took interest in the issue, the UNFCCC was the only multilateral climate change policy-making forum in which the country, like the other small GCC states, had participated. Change began around 2008 as new Abu Dhabi-based actors, including Masdar, started weighing in on the country's position-formation. The successful IRENA campaign elevated the issue to a new level and since 2010 the UAE, led by the Foreign Ministry, began sending subtle signs of a more balanced position to come.

UNFCCC-related decision-making

In common with many other countries in the Middle East, because of the lack of prior interest and domestic capacity, only in the mid-2000s did the UAE establish the relevant national UNFCCC-related institutions and begin preparing the documents required from non-Annex I parties.[70] The Designated National Authority (DNA) for the CDM, a precondition for participation in the mechanism, was established around 2006, with the help of the EAD and Masdar, the latter of which played an instrumental role in the activation of CDM projects in Abu Dhabi.[71] Reflecting the UAE's 'culture of committees', the DNA consists of two organs: the National Higher Permanent Committee for the CDM, presided over by the Ministry of Energy, and the CDM Executive Committee, the implementing organ, headed by the EAD. Both include local and federal level members, but are heavily Abu Dhabi-weighted.[72] In 2006, after many years of delays, partly due to data availability, a committee led by the Ministry of Energy presented the UAE's initial national communication to the UNFCCC. Although this 'Kyoto Committee' had been established in 2000, the process only gained momentum after the EAD hired a US-based research institute to take over the task.[73] The second communication from 2010 was still officially produced under the Ministry of Energy's coordination. The newest committee, the National Climate Change Committee, was established in 2010 'to set clear direction for climate policy at a national level'.[74] It has been headed by the Minister of Environment, after a brief period during which it was unclear whether the (more capable) new Ministry of Foreign Affairs-based DECC would take charge. The committee includes representatives from the Ministries of Energy, Environment and Water, and Economy, Abu Dhabi's Executive Affairs Authority and the Dubai Municipality, among others.[75] In general, stakeholders have noted an inter-emirate rivalry since the committee's establishment, personified in the Minister of Environment, under the patronage of Dubai's ruler Sheikh Mohammed, and DECC's head Sultan al-Jaber, under the patronage of Abu Dhabi's Crown Prince Sheikh Mohammed and Foreign Minister Sheikh Abdullah.

The UAE's policy in the UNFCCC, 1996–2011

Reflecting a pattern common for OPEC member states, the UAE's delegation has always had representatives from the oil sector. Until 2010, when the Ministries of Foreign Affairs and Environment and Water took over, the Ministry of Energy held the title of the National Focal Point.[76] However, the number of participants from other federal level institutions has almost always outnumbered the number of oil sector representatives. Abu Dhabi has generally been the only emirate with local level representation, including the Supreme Petroleum Council, the ERWDA/EAD and, since the late 2000s, Masdar (although not as part of the official negotiating team).[77] The size of the UAE Conference of the Parties (COP) delegations has been proportional to the country's size and developing country status, only growing after 2009.[78] In terms of holding chair positions or hosting UNFCCC workshops, the UAE's record reflects its relatively low-profile participation in the regime.[79] Remarkably, there are a few individuals who have remained on the lists of participants since the late 1990s or early 2000s until the present.

From the negotiating summary archives of the International Institute for Sustainable Development (IISD) and available high-level speeches from past conferences of parties, an image of a passive observer emerges: the country primarily appears supporting statements by OPEC member states, most prominently Saudi Arabia. Most positions expressed or supported by the UAE are identifiable with the following OPEC themes: uncertainty of climate science and obstructionism (1996); avoiding additional commitments vis-à-vis the convention (1997–98, 2009); impacts of response measures (1997–2010); opposing discussion of aviation and maritime transport in the UNFCCC (2005); and supporting CCS under the CDM (2009–10).[80]

More broadly, the UAE's interests in the negotiations are those of an 'average' developing country: in an interview in 2008, Environment Minister Rashid bin Fahad described the UAE's position as supporting the Bali Action Plan (a two-year roadmap agreed upon at the Bali conference in 2007), calling for Annex I to fulfil their obligations, and supporting the CCS/CDM issue. At the time he felt that the UAE would not be ready to take on any new obligations under a new post-2012 pact.[81] In 2009—apart from response measures, CCS, and additional commitments—three UAE negotiators who were interviewed described the country's main interests in the negotiations as continuation of the Kyoto Protocol, and technology transfer, finance and capacity building from the developed countries. Sharing the sentiment of most developing countries, the negotiators were disappointed that the developed countries lacked leadership and provision of support for the new commitments they were asking from developing countries. At the time, a stronger differentiation of interests among the different domestic stakeholder organizations was evident. It reflected the new domestic priorities emerging under Sheikh Mohammed bin Zayed's 'new energy economy': while the Energy Ministry's lead negotiator noted that 'everything the UAE does is for the protection of oil', an EAD negotiator implied that in the response measures issue, the UAE was re-evaluating its existing practice of following the OPEC group—a statement that showed the careful balancing act the UAE Ministry of Foreign Affairs had to work out between

the country's old reference group and new domestic priorities. One negotiator even suggested that domestic action without binding commitments, and even with binding ones if these were to be well supported by
the developed countries, would be possible.[82]

Notably, despite the central role of the Ministry of Energy in coordi-
nating the UAE's activities in relation to the UNFCCC, the Conference
of the Parties' ministerial speeches have generally been delivered by other
ministers and dignitaries.[83] Starting from 2007, the portrayal of the UAE
as a proactive participant in the regime emerged as a new theme in the
speeches. In Copenhagen in 2009, the Environment Minister Rashid bin
Fahad, in a show of solidarity, even mentioned a federal renewable energy
target of 7.5%,[84] which lamentably appears to have been dropped as a
result of inability to align policy goals across the federation.[85]

The year 2010 marked a new chapter in the UAE's engagement with
the UNFCCC, as the Ministry of Foreign Affairs (MoFA) promptly
took over the agenda and began consciously but prudently distancing
itself from Saudi Arabia.[86] In February 2010, persuaded by the US and
against the preferences of the Saudis,[87] the UAE associated itself with
the Copenhagen Accord as the first OPEC state (later to be followed
only by Algeria, Nigeria and Angola). In the association letter, the Minister of State for Foreign Affairs, Anwar Gargash, highlighted mainly
familiar issues: the historical responsibility of Annex I countries, the right
to development of developing countries, and the need to minimize the
adverse impact of response measures. As a new issue—obviously arising
from the domestic context—the letter mentioned nuclear energy, in addition to CCS, in connection with flexibility mechanisms (CDM).[88]
According to a close observer, interviewed in late 2010, the UAE now
wanted to be seen as constructive and contributing something important to the negotiations. While the MoFA unit was left to wait for more
substantial policy guidance from the national climate change committee
chaired by the Minister of Environment, it had already begun proactively
engaging in the negotiations in more technical areas.[89]

In December 2010 the Foreign Ministry's—and Abu Dhabi's—*de facto*
dominion over the country's external climate policy was publicly confirmed as Sheikh Abdullah delivered the UAE's high level speech, instead
of the Environment Minister, also present. Reflecting the mandate of his
ministry, Sheikh Abdullah emphasized, alongside domestic mitigation
actions, the country's international engagement in mitigation and adap-

tation, including (Masdar's) investments in renewables in Europe and pledges of US$350m in renewable energy projects in developing countries and support for small island states.[90]

Despite a promising start, the pitfalls of autocratic decision-making in a loose confederation, combined with high dependency on external rent, were apparent in the weak rooting of both the coordination of domestic climate change policies and Abu Dhabi's Masdar by 2011–12. As a sign of the ongoing lack of federal coordination, the UAE did not deliver a high-level speech in the 2011 COP in Durban, South Africa. The 2012's World Future Energy Summit was in turn accompanied by a disheartening statement by Masdar's CEO al-Jaber that Abu Dhabi's 7% renewables goal would be pushed back from 2020 to 2030, only three years after the goal had been announced in the very same event.[91]

In conclusion, the principles and main aims of the UAE towards the UNFCCC, characterized by a largely passive approach and greatly influenced by Saudi Arabia and the OPEC group, remained impressively constant during the period from 1995 to 2009. The new green forces in Abu Dhabi, filtering into the UAE's 'high-level position' in the late 2000s, wished to demonstrate to the international community a concern for climate change and a willingness to do one's equitable share. Policy alignments soon followed suit: in 2011, the UAE still continued to support in the negotiations the long-term positions of its closest reference group, OPEC, while already participating in the meetings of the Cartagena Dialogue of roughly thirty 'middle ground' countries. In 2012, the country was already seen visibly distancing itself from a new group of conservative 'like-minded' developing countries, under which most Arab and OPEC states had regrouped. While the future of the UAE's new climate policy is still uncertain, past evidence suggests that as long as Abu Dhabi's green elite perceive that abundant international political capital will be clearly achievable from a constructive image, the progressive side of the balance is likely to prevail.

Towards broader participation and engagement

Fortunately, as an indication of longer-term sustenance for the UAE's domestic natural sustainability initiatives, the country has become involved in multilateral governance of the climate change era in numerous other ways. From around 2009, as a consequence of the international attention to Abu Dhabi's alternative energy and environmental sustainability initiatives, the UAE became involved in a number of high-level working groups, fora and events, and found new international friends. In June 2009, in recognition of Masdar's international reputation, Masdar's CEO al-Jaber was named as a member of the UN secretary general Ban Ki-moon's Advisory group on Energy and Climate Change.[92] A year later, in August 2010, the secretary general appointed Sheikh Abdullah to a High-Level Panel on Global Sustainability.[93]

After securing IRENA, Abu Dhabi quickly became the preferred location for a number of climate change-related events. These included two meetings of the US-initiated Clean Energy Ministerials in 2010–11. As a mini-victory over its larger GCC neighbour, the UAE was the only Arab country in the twenty-three-member group, which included all major economies.[94] The capital was the location of the World Renewable Energy Congress of 2010 and the 33rd session of the Intergovernmental Panel on Climate Change, among other gatherings. Masdar, not losing a single opportunity, had, in 2010, already labelled Abu Dhabi as 'a global centre for renewable energy partnership'.[95] In the 2012 Earth Summit in Rio de Janeiro, the UAE's presence did not go unnoticed either, as it was, along with Qatar, among the few smaller 'developing' countries to host a national pavilion in the sustainability mega event.

As a result of Sheikh Abdullah's and Masdar's dynamism, the UAE also engaged with unexpected countries: in December 2009, Abdullah signed a joint statement with the foreign ministers of Cape Verde, Costa Rica, Iceland, Singapore and Slovenia, self-described as 'small points of green reference' within those states' own regions. The declaration included words previously unheard from a GCC OPEC member state, including openly calling for a 2°C limit for global warming.[96] In 2010, in a highly symbolic move, Abu Dhabi invited the Maldives' President Mohamed Nasheed, one of the lead representatives of the small island states AOSIS group, to speak at Masdar's World Future Energy Summit.[97]

As expected, the UAE's newfound role has also begun attracting calls to do even more, one of these coming from a lead IPCC author who suggested in 2009 that 'the UAE should work with developing countries to frame what the obligations of developing countries should be'.[98] As stressed repeatedly by the UAE's UNFCCC delegates, while Abu Dhabi seems committed to raising the share of renewables in its domestic energy mix, committing to climate change mitigation at the international level is still seen as a boundary not to be carelessly crossed. However, it may only be a matter of time before the UAE takes another pioneering step in the region by registering its domestic emission targets as a Nationally Appropriate Mitigation Action under the UNFCCC.[99]

The case of Qatar in the UNFCCC: stealing the show

The GCC/OPEC has been the primary reference group in determining Qatar's policies and behaviour in the international climate regime, in an even more pronounced way than for the UAE. Qatar's small but active role in the negotiations and its somewhat more defined policies reflect how, until 2012, international climate politics were nationally a low-key issue, enabling the energy sector to dominate policy. Since its onset, Qatar's role centred around defending the country's oil income from the perceived threat of global mitigation, and bluntly rejecting any calls for domestic mitigation by referring to the country's current role as a major exporter of 'clean' energy, gas, to the world. In late 2011, in a victory that took many international observers by surprise, Doha won its bid to host the 18th Conference o of the Parties to the UNFCCC. The conference will be a major test for Qatar's reputation as a skilful mediator and an emerging player in the Arab world. Even more important, it will be a paradigmatic test of the ability of the Qatari government and society to receive external criticism, allow for some form of street protest and, in general, stand in the global spotlight and be held accountable for events and issues *within* its borders.

UNFCCC-related decision-making

Qatar acceded to the UN Climate Convention in 1996 and the Kyoto Protocol in 2005. Immediately afterwards, it set up its Designated National Authority (DNA), hosted by the environmental authority SCENR.[100] CDM-related expertise in Qatar too has been mainly concentrated within one institution, Qatar Petroleum, owing to the company's massive al-Shaheen flaring gas recovery project.[101] In 2007, linked to the project, QP established a CDM Department and a high-level QP CDM Committee, charged with establishing a Qatar Carbon Management Plan, which, if drafted, was never made public.[102] In October 2007, an eight-member national climate change committee (NCCC) was reestablished by decision of the Heir Apparent and SCENR chairman, Sheikh Tamim. It included members from the SCENR, Qatar Petroleum and the Office of the Heir Apparent, among others.[103] UN archives show that prior to this, the group had existed under the Ministry of Energy and Industry since at least 2002.[104] Since the Environment Ministry's establishment, the committee has been chaired by the relevant (vice) minister and reports to this ministry.[105]

Qatar's policy in the UNFCCC, 1996–2011

Typically for an OPEC state, Qatar's official delegations to the UNFCCC have had strong representation from Qatar Petroleum and the Ministry of Energy and Industry. However, unlike the UAE and Saudi Arabia, but as in Bahrain, Kuwait and Oman, the Qatari Ministry of Environment is officially the country's National Focal Point.[106] Pinpointing the dominant institutions in external climate policy-making is difficult owing to the opacity of the process and the elusiveness of most negotiators.[107] While Qatar's policy positions strongly indicate an oil sector-led policy, a negotiator interviewed in 2009 assured that no hierarchy exists in formulation of policies in the NCCC.[108] Official lists of participants and archival material of past negotiations point towards two phases: a more passive phase in the 1990s, when environmental authorities represented Qatar, and a different phase in the 2000s, when the energy sector entered the picture, activating the country's participation in the negotiations.

Qatar's notified delegation size has been roughly similar to that of the UAE, rising considerably after Qatar launched its bid to host COP18 in 2009. In 1999–2007 the Qatari delegation was led by the QP manager Mohammed Jassim al-Maslamani, since 2008 by the Environment Minister al-Midhadi. Like the UAE's, the Qatari delegation has included a number of long-term negotiators.[109]

Qatar's policy positions in the UNFCCC have been extremely stable, and have been spelled out in great detail in a number of submissions of views in 2001–11.[110] In 2001–02, Qatar actively sought a special status for natural gas in the regime.[111] One of the submissions, a detailed 20-page presentation on the economic and environmental benefits of natural gas in 'global energy decarbonisation strategies', illustrated the impact of Qatar's gas industry on the country's total greenhouse gas emissions, with an eye to exempting the country from responsibility for its high emissions.[112] The more recent submissions, from 2006, 2009 and 2011, set the country's position on a wide range of topics: support for CCS in the CDM (although it was noted that lack of capacity and know-how constitute barriers to hosting projects in Qatar); the crucial importance of the response measures issue and Qatar's related vulnerability; and the broad architecture of the post-2012 climate deal, and in particular the Bali Action Plan (which has formed the basis for the negotiations on long-term cooperative action under the UNFCCC).[113] In these submissions Qatar, like OPEC countries generally, objects to additional commitments for developing countries, a clear distinction between commitments of the two main groups, and fiercely opposes differentiation within the developing country group (particularly on a GDP or greenhouse gas per capita basis). It calls for equal treatment of all greenhouse gases (to divert attention from CO_2 and, hence, oil), opposes trade regulations to energy-intensive exports from developing countries, and shuns any numerical goals for emission reductions by 2020 or 2050.

In parallel, Qatar's frequent appearance—relative to the country's size—in the IISD negotiating summaries reinforces the image of a static policy, almost identical to that of Saudi Arabia and certainly in line with typical OPEC positions, natural gas constituting the only noticeable difference. The response measures issue is the most important issue for Qatar, if measured by the frequency of its being mentioned in IISD archives (1997–2011). After this come the CCS/CDM question and technology transfer more broadly (2004–2010), and burden sharing and developing country commitments (1998–2009).[114] Additionally, the country has opposed discussing international maritime and aviation emissions in the UNFCCC (2008–2010).[115] Also, some of Qatar's past positions are identifiable as obstructionism (1999; 2010–11), as defined above.[116] Outlining the country's position in 2009, a background paper by a Qatari negotiator outlines four central issues in the post-2012 negotiations: continuing active participation in order to ensure that the country's interests are protected; avoiding commitments; response measures; and the CCS/CDM issue, as well as the functioning of flexibility mechanisms (for example CDM) more broadly.[117]

The central points mentioned in high-level speeches from in the 2000s do not differ significantly, although the speeches from 2009 and 2011, delivered by the Environment Minister al-Midhadi, had a more proactive and less defensive tone.[118] Similarly to the UAE, at Cancún in 2010 al-Midhadi, reflecting Qatar's domestic power hierarchy, ceded the high-level speech to the Energy Minister Abdullah al-Attiyah. Al-Attiyah's speech carried a sceptical tone towards international mitigation and a defensive one towards Qatar's mitigation actions, noting that the country had already contributed 'more than any other country' through its clean energy exports and calling for the developed countries to fulfil their commitments. Outside the UNFCCC as well, al-Attiyah, the foster father of Qatar's LNG programme, has spoken in a critical tone of alternative fuels and called for compensation for response measure-induced revenue losses.[119] Speeches from recent years imply that his scepticism about alternative energies[120] stems from indignation over the scapegoating of oil and gas producers and a worry for the future of natural gas demand.[121] The fact that al-Attiyah belongs to an older generation than, for example, Abu Dhabi's Sheikhs Mohammed and Abdullah bin Zayed may also have played a role in forming a world-view less amenable to the dangers of climate change and the possibilities of the low-carbon energy economy.

Remarkably, there has invariably been a more positive tone in Emir Hamad bin Khalifa Al Thani's comments in international fora where, in the late 2000s, he noted the need for cooperation and contribution by both developed and developing countries, described climate change as a serious threat, and spoke about the potential of solar energy.[122] Also, al-Midhadi has stressed the business opportunities for Qatar provided by low-carbon technologies, and the GSDP's National Vision 2030 document pledges 'support for international efforts to mitigate the effects of climate change'.[123]

In conclusion, the constancy of Qatar's positions in the international climate regime up to 2011 is striking. Despite the Emir's broader vision of sustainability, embodied in the Qatar National Vision, the international climate change regime had not received due attention from the country's natural sustainability-minded sectors and patrons. The strong hold of the energy sector on the country's external climate policy and its long-term policy of alignment with Saudi Arabia (and Kuwait) on most issues[124] resulted in weak representation of broader or opportunity-focused interests. During the observed decade and a half, the conservation of fossil fuel revenues remained the central theme and aim of Qatar's UNFCCC policy.

Towards a new era in Qatari climate politics?

As climate change and environmental issues became increasingly prom-inent in international affairs, the state became interested in hosting major energy and environmental events too. This aligned with the Qatar gov-ernment's goals of making the country a regional 'mini-superpower' in the areas of diplomatic mediation (and as the supporter of the Arab Spring in the Arab republics) and branding it as a renowned venue for international meetings. Since the establishment of the Ministry of Envi-ronment it had already hosted the Vienna Convention and Montreal Protocol on Substances that Deplete the Ozone Layer (COP8/MOP20) in 2008, and the Convention on International Trade in Endangered Spe-cies (COP15) in 2010. At the Copenhagen climate change conference in 2009, in an unsuspected move, Qatar offered to host COP18/CMP8 of the UNFCCC in 2012.[125] In an attempt to raise the profile of envi-ronmental issues in Qatar, arguably linked to the COP bid, the Minis-try of Environment organized in 2011 the Qatar International Environ-ment Protection Expo, ecoQ.

When Qatar signed up to the contest, it faced a tough competitor, South Korea, described in Western media as 'an emerging leader in the global quest to reduce greenhouse gas emissions' and 'a vocal supporter of low carbon economic models', which has in the past years announced major green investments, emission cuts and plans for a carbon trading scheme.[126] Those on Qatar's side pointed out that engaging with the OPEC countries was an important gesture, and might help to raise the issue on the GCC exporters' domestic agendas. Heavy lobbying was nat-urally involved, and high-level sources familiar with negotiations have suggested that Qatari campaigners employed both pressure and incen-tives to gain support.[127] Nevertheless, two years later, by COP17, held in November 2011 in Durban, neither of the contestants had backed away. According to observers present, in an astute show of negotiating skills Qatar gave the Asia group (which would decide on the host) two options, Doha or Bonn, while knowing that the group would regard choice of the UNFCCC secretariat's home city Bonn as a defeat.[128] The winner had emerged. Speaking to COP17, al-Midhadi ensured that the government would make every effort to make COP18 a success.[129] In May 2012, in what could be interpreted as mending fences with the first runner up, Qatar joined the consortium of founding members of South Korea's Global Green Growth Institute initiative.[130]

To judge from all available evidence, Qatar's bid stems from the country's broader external profile-building as a host for major global events. Run by the Foreign Ministry, the campaign idea bore few links to either the country's previous UNFCCC policies or Qatar's domestic natural sustainability initiatives and key players in the field. This is evident, for example, in the way the government did not appear to see any potential image conflict in the fact that it hosted in 2011 the 20th World Petroleum Congress, in the very same building that would exactly a year later convene some of the world's most aggressive opponents of fossil fuels, in the form of thousands of NGO observers at the COP18 conference. Qatar's candidacy also indicates rather clever farsightedness, as any new major treaty sealed in the conference would bear Doha's name, as happened in the current round of WTO talks. By 2011, with the level of global willingness to commit to emission cuts still dragging far behind that recommended by climate science, it had become realistic to expect that no major breakthrough was in sight for another few years. Still, Qatar had already scored image victories in many other areas, as the conference will be the first one to be organized in the Gulf region and in an OPEC member state.

In early 2012, with exceptionally little time left to prepare the ground for the conference, the magnitude of the endeavour began to dawn on the Qatari hosts. By January 2012, the government had set up a high-level national task force for the organizational, logistical and political aspects of the conference. The task force was chaired by former Energy Minister Abdullah al-Attiyah, co-chaired by the Minister of State in the Ministry of Foreign Affairs Khaled bin Mohammed al-Attiyah, and with three key ministers as members: the Energy and Industry Minister Mohammed al-Sada, the Environment Minister Abdullah al-Midhadi, and the chairman of the QNFSP, Khaled's brother Fahad bin Mohammed al-Attiyah.

For Qatar, COP18 will be a test and possibly a paradigm-changer, in many ways. Not only will the country be judged, as any host country would be, by the political and organizational successes or failures of the conference. Qatar will also become the centre of attention of global media and international NGOs of all possible kinds, with regard both to its ability to be seen as a balanced, fair facilitator and to its environmental record. Qatar's promising natural sustainability plans, developments and projects will receive an unprecedented PR opportunity, but the government will have to find a careful balance between reality and promises. Success will also be measured in the awareness-raising impact of the conference on Qatar's society. COP18 will also be a paradigmatic event in the sense that it will bring street protests—however small, peaceful or confined to restricted areas—straight to the country's capital. It will also test the government's and local society's ability to deal with external criticism, directed both at issues related to climate change, like per capita emissions, and at those completely unrelated, such as human rights or gender issues. COP18 may well become the first time that Qatar's domestic issues cannot be hidden behind Al Jazeera and the country's bold foreign policy. But most important, hosting the event will push Qatar to finally devise an external climate policy, anchored in the now diverse domestic interests vis-à-vis the issue.

9

CONCLUSIONS

Since the late 2000s, the Gulf monarchies have been confronted by multiple natural resource and environmental pressures, emanating from the domestic and external environment. These pressures, which have been examined in this book through the concept of natural unsustainability and the issue of climate change, threaten both the local environment and the stability of the existing social contracts. After an extensive examination of the Gulf monarchies' unsustainable natural resource consumption patterns and related environmental impacts of the past decade, and their dual vulnerability with regard to climate change, the book turned to two sets of questions: how have two structurally similar Gulf monarchies, Abu Dhabi and Qatar, responded to these challenges, and what drivers have influenced developments in each one? How have these monarchies dealt with the issue of climate change at the international level, and how have the evolving domestic energy and environmental agendas influenced each case?

Abu Dhabi's responses

Abu Dhabi's status as a member and leader of the seven-emirate confederation of the United Arab Emirates makes it a complex case study: despite its high level of independence in many areas of economy and policy-making, there are inter-emirate linkages, dependencies, interest conflicts and other contending aspects that needed to be included in the analysis. Abu Dhabi's heavy dominance of the country's oil and natural gas reserves (94% and 93%) and its wealth nevertheless sets it clearly apart from the other emirates. The emirate's oil and sovereign wealth-based affluence allows it to plan big, take its time, and even make some mistakes along the way. In the mid-2000s, led by the dynamic Crown Prince Sheikh Mohammed, Abu Dhabi embarked on an ambitious economic diversification endeavour, spearheaded by the development vehicle Mubadala, with the aim of creating a number of new high-value economic sectors, including alternative energies.

Towards the late 2000s, the UAE's economy and population were booming as a result of high oil prices. Abu Dhabi realized it was facing a looming gas shortage, while the poorer emirates, which it was supporting through both budgetary and energy allocations, found themselves amidst an energy crisis. Domestic sour gas reserves had not been developed in time and the high opportunity cost of using oil domestically made it an extremely unattractive option. Simultaneously, Abu Dhabi was securing increasing financial resources which would enable it to diversify its energy mix: between 2006 and 2008 the strategy became defined with the launching of two massive initiatives. These consisted of nuclear energy (in the medium term) and some renewable energy capacity and the aim to master alternative energy technologies (in the longer term). Unlike Bahrain and Oman, where fossil fuel-derived wealth was more limited, all Abu Dhabi needed for deploying its ambitious alternative energy strategies was political will and new technology. In the case of nuclear technology this implied the political support of key global suppliers, most importantly the United States. In the case of other alternative energy technologies, it required either massive investments in companies abroad or a strategy to attract foreign direct investment into Abu Dhabi.

First born from this context was Masdar, established in 2006. Its remit and budget quickly grew as the positive international attention increased its importance in the eyes of Sheikh Mohammed. The company's stated aim was nothing less than to transform the emirate into an energy technology exporter. In 2008, Abu Dhabi's government pledged it would support the company with US$15bn.

Masdar also took up a broader task, motivated by a wish to be a regional pioneer but also to win prestige and fame: to show the world through its 50,000-inhabitant eco utopia city project and domestic solar energy projects that environmental sustainability would be possible in one of the world's least sustainable places. Despite a spectacular start, the economic crisis of 2008 revealed a number of problems. Masdar's biggest early mistakes included the haste at which it publicized extremely ambitious targets, locked its technology choices, and rushed into implementation without proper feasibility studies. Masdar also lacked strategic clarity on whether it was a commercial project or a transformational vehicle of the government. Finally, the social and economic sustainability of Masdar City were not well thought out in the early plans. All these problems also reflected Abu Dhabi's broader sustainability challenges, namely expectations of quick profits and boom-inspired excesses in real estate investments, typical of a rentier state, and the blurred lines between public and private interest, typical for the monarchical regimes of the Gulf. Nevertheless, in 2012, Masdar still stands as the single most ambitious and successful cluster of alternative energy and technology investments and deployment in the Arab Middle East.

As a positive result of the late 2000s' convergence of domestic energy security issues, high oil prices, Masdar, and the relative strength of the Abu Dhabi's local environment agency, the government began paying heightened attention to broader environmental concerns. This brought environmental sustainability onto the emirate's strategic planning agenda in a comprehensive manner, unforeseen elsewhere in the region. The most important dynamic factor in bringing forth these responses was top level patronage. Another key factor was the government's invention of itself as the regional 'green energy leader'. This was closely in line with the elite's neotraditional environmental legitimacy mechanism, personified most prominently through 'the legacy of Sheikh Zayed'.

Notably, Sheikh Zayed also left a more a tangible green legacy, the Environment Agency—Abu Dhabi (EAD, established by Zayed as ERWDA), currently the strongest environmental institution in the GCC. Owing largely to the Crown Prince's attention to the issue and support to the EAD, Abu Dhabi's environmental sustainability problems and the potential negative impacts of climate change began receiving due attention. The sensitivity of both Abu Dhabi's and Dubai's leaderships to external criticism—regarding the UAE's environmental performance and CO_2 emissions—was something that distinguished the two from the other small Gulf monarchies, and also acted as an important catalyst to action.

The 'greening' of Abu Dhabi was thus a two-dimensional legitimacy quest, an attempt to please both external and domestic audiences. Towards

Arab and Western audiences Masdar was a prestige tool, aimed at attracting envy and admiration of the modernity and progressiveness of the government—true small state branding. The Masdar brand cleverly labelled Abu Dhabi as the environmentally sustainable oil producer. Although certainly influenced by external image considerations, this greening project was not a result of Western pressure.

Legitimacy-seeking in relation to domestic audiences was pre-emptive rather than reflective of values prevalent in the Emirati society, still relatively uninterested in and uninformed about the consequences of the UAE's high natural resources consumption rates and environmental deterioration. Through top-down awareness raising and other soft measures that would not rock the rentier bargain and repel expatriates, the green forces of Abu Dhabi sought to address the soaring resource consumption. Simultaneously, the leadership was able to portray itself as a visionary leader to the fast-transforming and young Emirati society where calls for attention to climate change and other environmental issues are bound to appear, sooner rather than later.

A parallel development was Abu Dhabi's nuclear energy programme. Unlike solar energy, nuclear came to be regarded by the government as a cost-effective way to enhance domestic energy security and diversify the energy palette. The prestige associated with nuclear technology undoubtedly increased the attractiveness of the option, but Abu Dhabi had to simultaneously convince its neighbours and the international community of its peaceful intentions. Set up with the assistance of the emirate's key external security allies, the United States and France, the nuclear programme became a national priority from 2007, dealt at the highest level by Emir Sheikh Khalifa and also the Foreign Minister, Sheikh Abdullah. Despite a similar time frame to that of Masdar City, international cycles of oil or real estate prices did not seem to affect the nuclear energy programme's implementation, which remained motivated by rising domestic energy demand projections. Hence, it became clear that the nuclear energy programme's primary purpose is to provide energy security, while that of Masdar will be the transfer of technology and know-how in other areas of alternative energy.

Apart from its ability to attract international investment, Masdar's domestic standing has been highly dependent on its international image and credibility. An important positive recognition was Abu Dhabi's victory in its campaign to host the IRENA headquarters, led by the Foreign Minister. With both Masdar and IRENA on its soil, siding with a group famous for its problematic negotiating tactics and even obstructionism in the international climate negotiations under the UNFCCC became an impending image issue, quickly picked up by the Minister. The UAE's association with the Copenhagen Accord, the new Directorate of Energy and Climate Change, and Sheikh Abdullah's participation in the Cancún climate conference were all rapid and momentous changes for a country that had until early 2010 slumbered in OPEC's shade. Slowed by the need to balance the interests of the oil and new energy economies, on the one hand, and competition between Abu Dhabi and Dubai, on the other, the UAE's position is steadily moving towards measured but still bold engagement with the opportunities of the international climate regime and green growth.

Qatar's responses

A unitary political system, with a small, 1.7-million population and high concentration of power among a few individuals, the dynamics and structures of Qatar's energy and environment-related decision-making are easier to grasp despite their higher opaqueness. As in Abu Dhabi, Qatar's economic strategy consists of fossil fuel rent-based growth and diversification into industrial sectors of comparative advantage, as well as a number of carefully selected non-oil sectors. Qatar's key goals, under the patronage of the Emir's wife Sheikha Mozah, include fostering a knowledge economy through high-profile foreign partnerships in education; and research and development into areas defined as national priorities. Simultaneously, Qatar has been successful in replacing rent from its oil reserves, which are estimated to last for four to five decades, with natural gas revenues, becoming in 2006 the world's largest LNG exporter. By diversifying its gas exports on a geographical scale, through different types of exports, and along the value chain, Qatar has built a robust strategy for securing a stable flow of continued external rent for the coming decades.

While natural gas for Qatar has been a catalyst for economic growth and development, its abundance has slowed down investments and badly needed improvements in the area of natural sustainability. Following a subregional trend, the government has explored nuclear and solar energy, but by 2012 there were no concrete plans on large-scale implementation.

The food security programme, under the Heir Apparent Sheikh Tamim, contained bold plans for sustainable domestic agriculture and clean energy, but these were still at a very early stage. Amidst the accelerating construction boom, sustainable building was emerging as a bottom-up trend in the construction industry (whereas in Abu Dhabi it emerged top-down).

The massive development and real estate projects of the 2000s, fomented by and fomenting the economic growth, created a vicious cycle of unexpectedly fast population growth and, consequently, domestic natural resource consumption. This prompted a strategic re-evaluation of Qatar's development priorities and its pace, culminating in 2008 in the Qatar National Vision 2030, which placed emphasis on the three pillars of sustainable development. A similar, yet more specific and diversification-oriented development strategy was published by Abu Dhabi the same year.

At the turn of the 2010s, owing to its natural gas abundance and successful timing in developing its easily exploitable gas fields, Qatar was in a very different domestic energy security situation from any other Gulf monarchy. Despite signs of an eventual impact on Qatar's export capacity from growing industrial and residential energy demand, the government could still afford to postpone addressing the inefficiencies of the domestic demand side. Because they lacked a strong economic impetus, government attempts to design and implement plans specifically aimed at addressing the high natural resource consumption and greenhouse gas emissions were half-hearted, at most.

The lack of a top-level elite patron for the issue also contributed significantly to the absence of natural sustainability-related measures. Evidence from Qatar's education sector demonstrates the large quantities of financial and human resources and political attention that the state can draw together for a strategic goal, if this is perceived to be in the personal interests of a leading figure. Qatar's most influential elite members, however, were occupied by their respective areas of authority and patronage. While sustainable development constituted the new umbrella for strategic thinking among the Qatari elite, the economic pillar was still leading the way. Three elite members whose respective areas were tangential with alternative energies and technologies were the Energy Minister (until 2011) Abdullah al-Attiyah, Sheikha Mozah, and Sheikh Tamim.

There was no strong institutional leader in this area either, nor did one emerge. The new local environmental institutions (first SCENR, then the Ministry of Environment), despite increasingly significant staff numbers, were not bestowed with strong leaders or mandate. They lacked capacity and clout to devise and implement society-wide environmental policies and deal with contemporary climate change-related questions. As a consequence, Qatar had a 'second sphere' of environmental governance, consisting of the GSDP and the Ministry of Energy/Qatar Petroleum, and even 'private' developers, each with specific interests and roles.

Arguably, al-Attiyah's personal scepticism towards the viability of renewable energies and his heavy fossil fuel portfolio meant that alternative energies or mitigation policies did not figure as priorities in Qatar's energy sector. Since energy policy was tightly in the court of the Ministry of Energy and Industry, and education and research in Sheikha Mozah's, all activities under the Qatar Foundation were bound to be scientifically oriented and primarily linked to knowledge-society building. The mandate of the Qatar Science and Technology Park, inaugurated in 2009 with Sheikha Mozah's support, was rather similar to that of Masdar, but focused on a wider range of fields, from a narrower, mostly R&D perspective. Its mandate included economic diversification, job creation, and contributing to Qatar's post-carbon economy. Renewable energy as a theme was taken forward also by Heir Apparent Tamim bin Hamad, under whom the Qatar National Food Security Programme, directed by Chairman Fahad al-Attiyah, began in 2009, exploring the impossible-sounding idea of enhancing arid Qatar's food security through solar desalination.

For a full decade, starting in the early 2000s, Qatar's attitude vis-à-vis the international climate regime was defined by a fossil fuel sector-heavy interest representation and the view that Qatar, as a major gas exporter but a tiny state, should not be held accountable for its high domestic emissions. As it was a slightly more active participant than the UAE, with more clearly defined positions on a number of issues, Qatar's policy remained impressively static throughout the 1990s and 2000s, and even despite the establishment of the Ministry of Environment in 2008, external climate policy continued to be run from the Ministry of Energy. This only began to change when the more powerful Ministry of Foreign Affairs included the hosting of a major UN climate change conference in its array of diplomatic endeavours.

From a comparative perspective, it is important to distinguish between Abu Dhabi's green energy leadership pursuit and Qatar's piecemeal approach to the issue. Both represented active, early Gulf oil exporter responses to domestic natural unsustainabilities and the transforming global energy agenda. But each was the result of a completely distinct domestic context, consisting of diverging energy security situations and diversification priorities, and differences in elite size, dynamics, and historic personalities. Other distinguishing factors included the local institutional setting, such as differences in the strength of environmental institutions and the locus of power in decision-making related to climate change. A shared feature of Masdar and the QSTP was their dependence on direct elite patronage. In Abu Dhabi, this support was initially only linked to Crown Prince Sheikh Mohammed, but later became embedded in a number of the emirate's strategic objectives, international promises and domestic imperatives, receiving broader elite support. In Qatar, in turn, Qatar Foundation's continuity is still somewhat less certain, the mandate and scope of its QSTP remains narrow, and the future success of the QNFSP's massive food security plans are still shadowed by the uncertainties typical of large GCC development plans.

In Abu Dhabi's case, the new domestic priorities that emerged through its new alternative energy initiatives had a marked influence on its external climate policy formulation. In Qatar, the diversity of emerging domestic interests was only starting to become conveyed at the foreign policy level, as a result of Qatar's victory in its bid to host the 18th Conference of the Parties to the UNFCCC in 2012. In this case, the need to formulate a more balanced national position emerged from a different direction, but it is expected to have a similar outcome to that in the UAE: a more balanced position in relation to the global challenge of climate change.

A future for naturally sustainable rentier monarchies?

The final remaining question is: can the Gulf monarchies ever tackle their natural unsustainabilities and become 'green'? In other words, are the existing rentier structures and ruling bargains, characterized by subsidized natural resources and authoritarian, opaque decision-making patterns, compatible with natural sustainability? Or will disbanding these structural and dynamic sources of unsustainability, through gradual reform of the subsidy regimes and liberalization of the political system, be the only possible route to truly sustainable development?

There are a number of areas in which the GCC states can work and have begun working to achieve a naturally more sustainable development trajectory, which addresses the challenges of climate change. Among the most important of these are energy source diversification, economic diversification, energy efficiency, and technology and knowledge transfer. In the area of energy source diversification, associated benefits will include domestic energy security, lower emissions, ability to export higher value fossil fuels, and, in the future, export revenues from alternative energy sources—once technologies become cheap enough to compete with fossil fuels. Diversifying to natural gas exports (for those monarchies that have suitable reserves), instead of oil, is also a better option, from both climate change and rentier wealth continuity perspectives, given the lower carbon intensity of gas and its growing role as a transitional fuel in the global low-carbon transformation. Economic diversification into non-oil productive sectors will also create jobs and make the economy more resilient to variations in oil prices. Energy efficiency and savings can create important economic savings, enhance domestic energy security, and produce important emission cuts. Rational use of water, too, can help cut energy use and emissions. While energy efficiency gains might also serve to temporarily postpone fuel and utility subsidy cuts, efficiency should nevertheless be seen in a broader perspective, enabling more sustainable use of resources in a carbon-constrained world. Sustainable building and planning can bring important energy savings too, and incentives and regulation in this area are badly needed. Climate change and natural sustainability are also sources of opportunities. Branding and image strategies in the areas of clean energy, green building and natural resource management, among others, can help attract foreign investment and technologies, and be used as a source of external and domestic prestige and legitimacy. Opportunities available through the international climate regime include technology transfer, carbon markets, and, again, international prestige and status.

Somewhat counter-intuitively, when evaluating the conditions for successful proactive responses to climate change and environmental sustainability to arise in the GCC, based on this study, the strength of the rentier state and elite autonomy emerge as the primary determinants. Despite the large inflows of external rent in the 2000s, the simultaneous growing demand and gradual depletion of domestic fossil fuel resources overstretched the ruling bargain in Bahrain and Oman, which were already moving towards being post-rentier states. As a result, the governments could not afford large investments in expensive and potentially risky long-term diversification ventures. Nor was there much international attention to these two states' greenhouse gas emissions or other environmental unsustainabilities. The five smaller emirates of the UAE were in a similar situation, although many of them never had fossil fuel resources to begin with. Dubai, a relatively diversified rentier state, demonstrated early interest in green building and cutting its ecological footprint, but the economic crash in 2008 brought many positive developments to a temporary halt. Because of its large population Saudi Arabia has less wealth per capita to allocate, but it is still classified as a rentier state. Owing to its sheer size, implementing existing alternative energy plans will inevitably take longer than in neighbouring Abu Dhabi, for example. But at the same time, owing to Saudi Arabia's large population and industries, and the high share of oil in its domestic energy mix, the economic case for alternative energies there is much stronger, portending major projects to come. In the case of Kuwait, in turn, the state's allocation capacity remained strong, but the political system, although it is the most democratized of the monarchies, was locked in a quasi-permanent state of tension and crisis. This meant that any alternative energy projects, like a nuclear energy programme, or major natural resource policies are likely to be preceded by long debates and, consequently, delays. The best placed, both from a natural sustainability and regime survival perspective, are Qatar and Abu Dhabi, the two remaining monarchies with large remaining fossil fuel resources, small national populations and autonomous elites. But even these strong rentier states will need to continue to invest the coming decades' fossil fuel revenues skilfully and patiently, if they wish to survive in the global low-carbon era.

Authoritarianism arguably has both positive and negative consequences for the Gulf monarchies' sustainability. As was shown by the case of Abu Dhabi's nuclear programme, concentration of power and the suppression of domestic political debate enabled a fast start to the implementation of four nuclear plants that will greatly enhance the energy security of the entire federation. On the other hand, the lack of freedom of speech poses a fundamental value dilemma: whether energy security considerations should precede democratic participation. Moreover, the arbitrary decision-making patterns, for example in environmental permitting in Qatar, and the lack of independent local environmental NGOs keeping a check on the major polluters and other sources of environmental threats in most GCC states, are among the main reasons why most sectors and actors in the Gulf monarchies, despite often claiming green credentials, have been able to continue their environmentally unsound practices, business as usual.

At the very root of the problem of natural unsustainability has been the lack of awareness of, and consequently indifference to, the environmental consequences of development so far, which run from top to bottom and back. The tendency of the past decades in the GCC has been to place economic and growth sustainability ahead of social and environmental sustainability. This is now fortunately changing. Just as there is no silver bullet solution to climate change, the major environmental challenge of our times, there is no silver bullet solution to the GCC states' natural sustainability. Several policy measures need to be taken.

First, environmental sustainability should be treated as a cross-sectoral topic instead of a single sector, under one ministry or agency. There are still major obstacles to inter-institutional coordination that should be addressed by giving strong, dynamic and progressive institutions strong enough mandates in this area. Political determination to push environmental issues on the development agenda must come from the top and rules and regulations must be enforced, as this is the only way to ensure the participation of all government institutions and fair treatment of domestic and foreign business interests. Secondly, a change in pricing patterns must ensue. This is a non-negotiable precondition for achieving rational natural resource consumption patterns. Different options exist for raising consumer prices while maintaining the ruling bargain, including monthly, fixed-sum payments for nationals. Encouraging an open debate on the environmental externalities of existing consumption patterns will also help pave the way. Thirdly, public awareness-raising on the environmental impact of Gulf lifestyles, particularly among the children and youth, must be made in a consistent manner, and should go beyond conservation issues and beach clean-up campaigns. The youth of the GCC, currently alienated from their natural desert environment by the rapid modernization and wealth, need to be brought back (out from their air-conditioned cocoons and tank-size SUVs) to nature, to learn to appreciate it. Gulf monarchies need to educate and engage in a dialogue with their national and expatriate populations about the harmful impacts of prevailing natural resource consumption patterns. In particular, the governments should find ways to bring the national populations on board in conserving energy and water, and leading environmentally more sustainable lifestyles more broadly. Adapting the famous saying about Romans and Rome, it is the Qataris, Emiratis, Saudis and so on, who set the example for all other residents of their respective countries. However, under the existing ruling bargains, and the 'rentier mentalities' these create, this might yet prove to be the most difficult task.

Fourthly, the GCC states need an enabling 'infrastructure for sustainability': urban planning, transport infrastructure and waste management should urgently be upgraded. The GCC states need extensive public transport systems that cater for all segments of society; functional recycling industries and collection infrastructure; and urban planning that encourages walking and cycling in the cooler months of the year. Feed-in-tariffs and other incentives for cleaner energy sources and practices are other essential elements of an enabling environment for sustainability. Fifthly, as already alluded, regulation and enforcement should be made transparent and universally applicable to all industries and companies. Lack of available data on air pollution, for example, may be a sign of hidden environmental problems. If oil and gas companies' environmental performance is up to standard, they should have nothing to conceal. Lack of data and accessibility, more generally, are also major hindrances to effective policy-making. Furthermore, as outlined above, green diversification agendas bear major opportunities for particularly the resource-richer GCC monarchies, and these should be actively and consistently pursued.

Finally, regional and international coordination and cooperation in climate change mitigation and adaptation are always beneficial and bring important synergies. Regionally, cooperation should focus at least on sharing of data and best practices, instead of competition and duplication. The GCC interconnection grid and even the nuclear initiative, to some extent, are positive examples. The numerous codes currently in use in the area of green building codes are a negative one. Internationally, the sooner the Gulf monarchies start constructively and actively participating in the international negotiations on climate change, the better their chances of being on the winners' side of the energy transition will be. The global economy will eventually become low carbon.

It is arguably unlikely that the allocative rentier monarchies of the Gulf, as we now know them, will be able to evolve into significantly greener societies. Although the deterministic logic of the rentier structures might ultimately render a fundamental transformation impossible, this study has demonstrated that agency can indeed win a battle. Abu

Dhabi's Masdar represents an innovative push away from the allocation state, as do some parts of Qatar Foundation and its technology development initiatives. Squandering of natural resources and environmental unsustainability are, nevertheless, likely to persist in the Gulf monarchies as long as the existing material ruling bargains do. In the coming decades, Gulf rentier monarchies, and their governments, are unlikely to survive without addressing the increasingly pressing structural environmental and natural resource-related unsustainabilities discussed in this study.

NOTES

INTRODUCTION

1. WWF, *Living Planet Report 2012* (2012), p. 43; World Resources Institute (WRI), *CAIT 8.0* (Washington, 2011).
2. See e.g.: Kristian Coates Ulrichsen, 'Internal and External Security in the Arab Gulf States', *Middle East Policy*, 16 (2009), p. 41.
3. E.g. Raymond Hinnebusch, *The International Politics of the Middle East* (Manchester University Press, 2003); Gerd Nonneman (ed.), *Analyzing Middle East Foreign Policies and the Relationship with Europe* (Abingdon, Oxon: Routledge, 2005).
4. Mari Luomi, 'Gulf of Interest: Why Oil Still Dominates Middle Eastern Climate Politics', *Journal of Arabian Studies*, 1 (2011), pp. 249–66.

1. THE GULF MONARCHIES AND NATURAL UNSUSTAINABILITY

1. International Energy Agency (IEA), *Energy Balances of Non-OECD Countries* (2011), p. II.357; WRI, *CAIT 8.0*.
2. Mohamed A. Dawoud, *Water Scarcity in GCC Countries: Challenges and Opportunities*, Gulf Research Center Research Paper (2007).
3. Eckart Woertz, 'The Gulf Food Import Dependence and Trade Restrictions of Agro Exporters in 2008', in S. Evenett (ed.), *Will Stabilisation Limit Protectionism? The 4th GTA Report* (London: Centre for Economic Policy Research, 2010), pp. 49–50.
4. On the Dubai model, see e.g. Martin Hvidt, 'The Dubai Model: An Outline of Key Development-Process Elements in Dubai', *International Journal of Middle East Studies*, 41 (2009), pp. 397–418.
5. *The Telegraph*, 20 August 2005; *The National*, 23 June 2009.
6. AMEinfo, 12 September 2011.
7. Peak electricity demand. DEWA website: http://www.dewa.gov.ae/, accessed in September 2011.

8. Gerald Butt, 'Oil and Gas in the UAE', in Ibrahim Al Abed and Peter Hellyer (eds), *United Arab Emirates: A New Perspective* (London: Trident Press, 2001), p. 237.

9. The IPCC suggested in its Fourth Assessment Report in 2007 that global emissions should be stabilised at 445–490 ppm of CO_2 equivalent (or 350–400 ppm of CO_2) in order to limit the global average temperature increase at 2.0–2.4°C from pre-industrial levels. This scenario assumes that global emissions peak by 2015 and that they are reduced by 50–85% from 2000 levels by 2050. Intergovernmental Panel on Climate Change (IPCC), *Climate Change 2007: Synthesis Report. Summary for Policymakers* (November 2007), p. 20.

10. Sharon Burke, *Natural Security*, CNAS Working Paper (2009).

11. World Commission on Environment and Development, *Our Common Future*, document A/42/427 (United Nations, 1987).

12. Matthew Gray, *A Theory of 'Late Rentierism' in the Arab States of the Gulf*, CIRS Occasional Paper (2011), p. 24.

13. Steffen Hertog, *Princes, Brokers, and Bureaucrats: Oil and the State in Saudi Arabia* (Ithaca, NY and London: Cornell University Press, 2010); Abdulkhaleq Abdullah, *Contemporary Sociopolitical Issues of the Arab Gulf Moment*, LSE Kuwait Programme Research Paper, No. 11 (2010).

14. Hossein Mahdavy, 'Patterns and Problems of Economic Development in Rentier States: the Case of Iran' in M. A. Cook (ed.), *Studies in the Economic History of the Middle East: From the Rise of Islam to the Present Day* (Oxford University Press, 1970), p. 428.

15. Hazem Beblawi, 'The Rentier State in the Arab World' in H. Beblawi and G. Luciani (eds), *The Rentier State* (New York: Croom Helm, 1987), p. 49.

16. Mahdavy, 'Patterns and Problems', pp. 466–7.

17. Beblawi, 'Rentier State', pp. 51–2.

18. Hazem Beblawi and Giacomo Luciani, 'Introduction', in H. Beblawi and G. Luciani (eds), *The Rentier State* (New York: Croom Helm, 1987), p. 13.

19. Data for latest available year. Fuel revenue and GDP: World Bank, *World Development Indicators & Global Development Finance*, online database, accessed in January 2012; Statistics Centre—Abu Dhabi (SCAD), *Abu Dhabi Statistical Yearbook—2011*, Economy Chapter (September 2011), pp. 19, 29, 53; International Monetary Fund (IMF), *United Arab Emirates: Selected Issues and Statistical Appendix*, IMF Country Report No. 11/112 (2011). Government revenue: May Khamis *et al.*, *Impact of the Global Financial Crisis on the Gulf Cooperation Council Countries and Challenges Ahead* (Washington: IMF, 2010), p. 5. Although Khamis *et al.* use 'oil revenue', Qatari sources suggest this figure includes gas.

20. Christopher Davidson, *The United Arab Emirates: A Study in Survival* (Boulder: Lynne Rienner, 2005), pp. 30–8.

21. Qatar National Bank, *Qatar—Economic Insight* (September 2011), p. 4.

22. Population and labour: Martin Baldwin-Edwards, *Labour Immigration and the Labour Markets in the GCC Countries: National Patterns and Trends*, LSE Kuwait

Programme Research Paper, No. 15 (2011), p. 11; SCAD, *Abu Dhabi Statistical Yearbooks—2010 and 2011*; Dubai Statistics Center, *Population Bulletin, Emirate of Dubai 2010* (2011); Economist Intelligence Unit (EIU), *United Arab Emirates: Country Profile* 2008 (2008), p. 12. GDP: World Bank, *World Development Indicators*, accessed in January 2012; IMF, *United Arab Emirates: Statistical Appendix*, IMF Country Report No. 09/120 (2009). Oil reserves: BP, *Statistical Review of World Energy* (2011); US Energy Information Administration (US EIA), *Bahrain Country Analysis Brief* (March 2011). Oil reserve data for Bahrain are for January 2011.

23. Gerd Nonneman, *Political Reform in the Gulf Monarchies*, CMEIS Working Paper, University of Durham (2006), pp. 31, 37.
24. See e.g.: Rex Brynen *et al.*, 'Introduction: Theoretical Perspectives on Arab Liberalization and Democratization', in R. Brynen *et al.* (eds), *Political Liberalization & Democratization in the Arab World: Theoretical Perspectives* (London: Lynne Rienner, 1995), p. 15.
25. Beblawi, 'The Rentier State', p. 52.
26. EIU, *Democracy Index 2010: Democracy in Retreat* (2010); Freedom House, 'Freedom in the World 2010', http://www.freedomhouse.org/, accessed in May 2011; Baldwin-Edwards, *Labour*, p. 15; SCAD, *Statistical Yearbook 2010*, p. 226.
27. Terry Lynn Karl, 'The Perils of the Petro-State: Reflections on the Paradox of Plenty', *Journal of International Affairs*, 53 (1999), passim; quotes from pp. 36–7. Karl defines as petro-states OPEC states and Mexico.
28. World Bank, *World Development Indicators*, accessed in September 2011.
29. Sovereign Wealth Fund Institute, 'Sovereign Wealth Fund Rankings', http://www.swfinstitute.org/, updated in September 2011.
30. Brad Setser and Rachel Ziemba, *GCC Sovereign Funds: Reversal of Fortune*. CFR Working Paper (2009), p. 2.
31. Correspondence with Jim Krane, February 2012.
32. Gregory Gause, 'The Persistence of Monarchy in the Arabian Peninsula: A Comparative Analysis', in Joseph Kostiner (ed.), *Middle East Monarchies: The Challenge of Modernity* (Boulder and London: Lynne Rienner, 2000), pp. 170–4.
33. E.g. Nonneman, *Reform*.
34. Lisa Anderson, 'Dynasts and Nationalists: Why Monarchies Survive', in Kostiner (ed.), *Monarchies*, p. 64.
35. Michael C. Hudson, *Arab Politics: The Search for Legitimacy* (New Haven and London: Yale University Press, 1977), *passim*.
36. Mehran Kamrava, 'Royal Factionalism and Political Liberalization in Qatar', *Middle East Journal*, 63 (2009), pp. 401–20.
37. Nonneman, *Reform*, p. 3.
38. Davidson, *Survival*, pp. 66–7, 70–87.
39. Christopher Davidson, 'Abu Dhabi's New Economy: Oil, Investment and Domestic Development', *Middle East Policy*, 16 (2009), p. 69.
40. January 2010 prices. UK Department of Energy and Climate Change, 'Energy

Price Statistics', http://www.decc.gov.uk/; US EIA, 'Retail Gasoline Historical Prices', http://www.eia.gov/, accessed in September 2011.

41. Food and Agriculture Organisation of the United Nations (FAO), *Aquastat. Country Fact Sheet: Kuwait*, (September 2011).

42. Reserve data: BP, *Statistical Review of World Energy* (2011); US EIA, *Bahrain Country Analysis Brief* (March 2011). Petrol prices: *Emirates 24/7*, 15 July 2010; US EIA, *United Arab Emirates Country Analysis Brief* (November 2009).

43. World Bank, *World Development Indicators*, accessed in September 2011. Abu Dhabi data (estimate, 2005–2010): SCAD, *Statistical Yearbook 2011*, p. 19.

44. Wes Harry, 'Employment Creation and Localisation: The Crucial Human Resource Issues for the GCC', *International Journal of Human Resource Management*, 18 (2007), pp. 134, 142.

45. World Bank, *World Development Indicators*, accessed in September 2011.

46. John Chalcraft, *Monarchy, Migration and Hegemony in the Arabian Peninsula*, LSE Kuwait Programme Research Paper No. 12 (2010), p. 26.

47. Harry, 'Employment', pp. 134–6.

48. Bahrain Economic Development Board, *From Regional Pioneer to Global Contender: The Economic Vision for 2030* (2008), p. 7; GSDP, *Qatar National Vision 2030* (2008), pp. 18, 29; Ministry of National Economy of Oman, *Second Long Term Development Strategy 1996–2020* (2008); Government of Abu Dhabi, *The Abu Dhabi Economic Vision 2030*, Context and Executive Summary (2008), pp. 5–7, 13; Government of Dubai, *Dubai Strategic Plan 2015: Highlights* (2007), p. 22.

49. *World Development Indicators*, accessed in September 2011.

50. *MEED*, 8 September 2006.

51. ESCWA cited in: Harry, 'Employment', pp. 134–6. Data for Kuwait and the UAE n/a. Female data only for Qatar. Governments have different definitions for 'youth', the most common being 15–24 years.

52. World Bank, *World Development Indicators*, accessed in September 2011.

53. Euromonitor quoted in: *Yahoo News Maktoob*, 13 April 2010. See also: Mouawiya Al Awad and Carole Chartouni, *Explaining the Decline in Fertility among Citizens of the G.C.C. Countries: the Case of the U.A.E.*, ISER Working Paper, No. 1 (2010).

54. Andrzej Kapiszewski, *Arab versus Asian Migrant Workers in the GCC Countries*, UN/POP/EGM/2006/02, UN Population Division (22 May 2006), p. 11.

55. Correspondence with Christopher Davidson, December 2010.

2. THE EMERGING DOMESTIC ENERGY INSECURITY

1. General Secretariat for Development Planning (GSDP), *Second National Human Development Report. Advancing Sustainable Development: Qatar National Vision 2030* (2009), p. 108.

2. Steffen Hertog and Giacomo Luciani, *Energy and Sustainability Policies in the GCC*, LSE Kuwait Programme Working Paper, No. 6 (2009), p. 6.

3. IEA, *Non-OECD and OECD Energy Balances* (2012).

4. Correspondence with Jim Krane, February 2012.

5. EIU, *The GCC in 2020: Resources for the Future* (2010), p. 5.

6. See e.g. Hertog and Luciani, *Sustainability*, pp. 5–6; Booz & Company, 'Gas Shortage in the GCC: How to Bridge the Gap', *Perspective* (2010), p. 3.

7. IEA, *Non-OECD Energy Balances* (2011); US EIA, *Oman Country Analysis Brief* (February 2011).

8. Booz & Company, 'Gas Shortage', p. 5.

9. Correspondence with Laura El-Katiri, October 2011. US EIA, 'International Electricity Price and Fuel Costs', http://www.eia.gov/, updated in June 2010. Data n/a for entire EU.

10. Hertog and Luciani, *Sustainability*, pp. 2, 6, 8. Booz & Company, 'Gas Shortage', 5.

11. OSEC, *Cleantech Business in the GCC: Market Assessment Report 2009* (2009), p. 35.

12. IEA, *Non-OECD Energy Balances* (2011); correspondence with Laura El-Katiri, October 2011. Kuwait electricity prices for 2007.

13. US EIA, *Qatar Country Analysis Brief* (December 2009).

14. *Reuters*, 22 April 2010.

15. Correspondence with Laura El-Katiri, January 2012; *Gulf Daily News*, 28 October 2010; *Bloomberg*, 30 November 2010.

16. OSEC, *Cleantech*, p. 35.

17. E.g. interview with Robin Mills, Abu Dhabi, December 2011; *Gulf News*, 31 May 2010.

18. *The National*, 6 January 2010; 9 May 2010; 29 July 2010; WAM, 2 March 2012.

19. IEA, *Non-OECD Energy Balances* (2011).

20. Ibid.

21. Ibid.

22. Hertog and Luciani, *Sustainability*, p. 2.

23. *MEED*, 16 November 2007.

24. Laura El-Katiri, *Interlinking the Arab Gulf: Opportunities and Challenges of GCC Electricity Market Cooperation*, Oxford Institute of Energy Studies, EL 8 (2011), p. 1; *The National*, 29 July 2010.

25. See: http://www.desertec.org/.

26. Sogreah, *GCC Countries—GCC Water Grid: Preliminary Feasibility Study*, undated brochure.

27. Imen Jeridi Bachellerie, *Renewable Energy Transition in the GCC: Finding the Right Paradigms*, GRC Analysis (2010); OSEC, *Cleantech*, p. 35; *Saudi Gazette*, 3 October 2011.

28. GTZ, 'Wind power project on Sir Bani Yas island', http://www.gtz.de/, accessed in December 2010.

29. *MEED* 28 March 2010; *The National*, 13 September 2010.

30. *Saudi Gazette*, 23 June 2011; *Bloomberg*, 11 May 2012.

31. *MEED*, 28 March 2010; *Arabian Business*, 10 January 2011.

32. *Gulf News*, 25 October 2008.

33. *Construction Week*, 17 January 2009.

34. See e.g.: Abbas Kadhim, 'The Future of Nuclear Weapons in the Middle East', *Nonproliferation Review*, 13 (2006), p. 586.

35. EIU, *Qatar: Energy Report*, EIU Industry Briefing (1 December 2009).

36. International Institute for Strategic Studies (IISS), *Nuclear Programmes in the Middle East: In the Shadow of Iran*, IISS Strategic Dossier (2008), pp. 36, 55. Parts of this and the following paragraph published in Mari Luomi, 'The Economic and Prestige Aspects of Abu Dhabi's Nuclear Programme', in Mehran Kamrava (ed.), *The Nuclear Question in the Middle East* (London: Hurst, 2012), pp. 125–58.

37. *Global Security Newswire*, 9 March 2008.

38. IISS, *Shadow*, p. 36.

39. Correspondence with Laura El-Katiri, January 2012.

40. Government of the UAE, *Policy of the United Arab Emirates on the Evaluation and Potential Development of Peaceful Nuclear Energy* (20 April 2008), p. 1.

41. *Financial Times*, 25 March 2009; *Gulf News*, 18 July 2008; *The National*, 6 January 2010.

42. *Bloomberg*, 9 March 2011; *Reuters*, 2 July 2010.

43. IEA, 'World Energy Outlook 2010: Presentation to the Press', http://www.worldenergyoutlook.org/docs/weo2010/weo2010_london_nov9.pdf, dated 9 November 2010.

44. Hertog and Luciani, *Sustainability*, pp. 2, 7.

45. Qatar Cool presentation, Qatar-Korea Renewable Energy Seminar, Doha, October 2011.

46. Organization of Petroleum Exporting Countries (OPEC), *World Oil Outlook 2010* (2010), p. 84.

47. *MEED*, 5–11 February 2010.

48. *Arabian Business*, 14 December 2010; *Reuters*, 4 October 2011.

49. The book uses a slightly different definition for 'old' and 'new economy' than Davidson in 'New Economy', p. 69.

50. Fédération Internationale de Football Association, *2022 FIFA World Cup Bid Evaluation Report: Qatar* (2010), p. 11.

51. IPCC, *Synthesis Report*, *passim*.

52. E.g. *The Daily Star*, 2 September 2009.

53. See e.g. Paul Aarts, *The Arab Oil Weapon: A One-Shot Edition*, Emirates Occasional Paper, No. 34 (ECSSR, 1999). OPEC's share of global oil supply is, however, projected to rise in the coming decades.

54. US EIA, *Bahrain Country Analysis Brief* (March 2008); US EIA, *Oman Country Analysis Brief* (April 2007); Oxford Business Group, *The Report: Dubai* (2007), p. 122; *Petroleum Economist*, 29 July 2010.

55. Anthony Giddens, *The Politics of Climate Change* (Cambridge: Polity Press, 2009), p. 44.

56. Matthew Paterson, *Global Warming and Global Politics* (London: Routledge, 1996), p. 1.
57. IPCC, *Synthesis Report*, p. 2
58. See e.g. European Commission, *Green Paper: A European Strategy for Sustainable, Competitive and Secure Energy*, COM (2006) 105 final (8 March 2006); World Wildlife Fund for Nature (WWF), *No Energy Security without Climate Security* (2006), p. 2.
59. Matthew Hulbert and Tariq Akbar, *Why a Gas Troika and Cartel Will Prove to be Hot Air*, Datamonitor Global Analysis (19 November 2008), pp. 4–5.
60. See Bassam Fattouh, *The Drivers of Oil Prices: The Usefulness and Limitations of Non-Structural Model, the Demand-Supply Framework and Informal Approaches*, Oxford Institute for Energy Studies, WPM 32 (2007).
61. OPEC, *World Oil Outlook 2008* (2008), p. 5.
62. *Financial Times*, 4 November 2008; *The National*, 10 November 2008.
63. IEA, *World Energy Outlook 2005: Middle East and North Africa Insights* (2005), pp. 125–6. Cordesman has gone even further to argue that the Iraq-Iran War also played a role, as the smaller Gulf states hoped that larger reserves would attract more external aid and boost their political status. Anthony H. Cordesman, *Energy Developments in the Middle East* (Westport: Praeger, 2004), p. 9.
64. BP, *Statistical Review of World Energy* (2011).
65. IEA, 'Presentation to the Press'; IEA, 'World Energy Outlook 2010 Fact Sheets', http://www.worldenergyoutlook.org/docs/weo2010/factsheets.pdf, dated 2010, quote from p. 2.
66. OPEC, *World Oil Outlook 2010*, pp. 46–53, 63.
67. IEA, 'Fact Sheets'.
68. OPEC, *World Oil Outlook 2010*, pp. 22–3.
69. John V. Mitchell and Paul Stevens, *Ending Dependence: Hard Choices for Oil-Exporting States*, Chatham House Report (2008), p. 20.
70. See e.g.: *Emirates Business 24/7*, 3 November 2010; *Reuters*, 11 December 2010; *Business Insider*, 14 June 2012.

3. FACING UNSUSTAINABILITY: THE CHALLENGES OF CLIMATE CHANGE

1. Some earlier parts of this subchapter were published in Luomi, 'Gulf of Interest'.
2. In its *World Energy Outlook 2008*, the IEA pointed out that even if all OECD countries cut greenhouse gas emissions to zero by 2030, this would not be enough to reach a safe level of global emissions. IEA press release, 12 November 2008.
3. UN General Assembly, *Declaration on the Right to Development*, A/RES/41/128 (4 December 1986).
4. The OAPEC member states are: Algeria, Bahrain, Egypt, Iraq, Kuwait, Libya, Qatar, Saudi Arabia, Syria, Tunisia and the United Arab Emirates.
5. 'Ambitious' generally refers to policies and measures aimed at limiting global temperature increase to 2°C.

6. WRI, *CAIT 8.0.*
7. Ali Hamed Al Mulla, 'Post 2012 Kyoto Protocol Climate Change Negotiations: Issues and Strategic Challenges to Qatar', paper presented at the *7th Natural Gas Conference*, Doha, 10–12 March 2009.
8. Paul Aarts and Dennis Janssen, 'Shades of Opinion: The Oil Exporting Countries and International Climate Politics', *The Review of International Affairs*, 3 (2003), p. 384.
9. Parts of this subchapter were published in Luomi, 'Gulf of Interest'.
10. IPCC, 'Summary for Policymakers', in M.L. Parry *et al.* (eds), *Climate Change 2007: Impacts, Adaptation and Vulnerability*, contribution of Working Group II to the Fourth Assessment Report of the Intergovernmental Panel on Climate Change (Cambridge University Press, 2007), pp. 13–16.
11. The term was first mentioned in CNA, *National Security and the Threat of Climate Change* (2007), p. 6. Since around 2007, climate change has been increasingly perceived as a security issue by scholars and decision-makers, especially in the Anglo-Saxon security community.
12. UN Security Council, 'Security Council holds first-ever debate on impact of climate change on peace, security, hearing over 50 speakers', SC/9000 (17 April 2007).
13. UN General Assembly, *Climate Change and Its Possible Security Implications*, A/63/L.8/Rev.1 (18 May 2009).
14. World Bank, 'Adaptation to climate change in the Middle East and North Africa region', http://go.worldbank.org/B0G53VPB00, accessed in December 2010.
15. Jessica Tuchman Mathews, 'Redefining Security', *Foreign Affairs*, 68 (1989), pp. 164–5.
16. See e.g. Mohamed A. Raouf, *Water Issues in the Gulf: Time for Action*, Middle East Institute Policy Brief, No. 22 (2009), p. 10.
17. See e.g. Mostafa K. Tolba and Najib W. Saab (eds), *Arab Environment: Future Challenges*, Arab Forum for Environment and Development (2008), pp. x–xi.
18. R.V. Cruz *et al.*, 'Chapter 10: Asia', in M.L. Parry *et al.* (eds), *Climate Change 2007: Impacts, Adaptation and Vulnerability* (Cambridge University Press, 2007), p. 480; UK Met Office, *Climate Change and the Middle East* (2009); Ministry of Energy of the UAE (MoE-UAE), *The United Arab Emirates: Second National Communications to the Conference of the Parties of UNFCCC* (2010), p. xiii.
19. IPCC, *Synthesis Report*, p. 8.
20. World Bank, 'A Strategy to Address Climate Change in the MENA Region', http://go.worldbank.org/OIZZFRJZZ0, dated 2 October 2008.
21. MoE-UAE, *The United Arab Emirates: Initial National Communication to the UNFCCC* (2006), pp. 36–7.
22. Raouf, *Water*, p. 1.
23. Oli Brown and Alec Crawford, *Rising Temperatures, Rising Tensions: Climate Change and the Risk of Violent Conflict in the Middle East*, International Institute for Sustainable Development (2009), pp. 8–18.

24. FAO website: http://www.fao.org/, accessed in January 2012.
25. 40% in Oman and 85–99% in the other monarchies in 2005. Dawoud, *Scarcity*.
26. Environment Agency—Abu Dhabi (EAD), 'Water resources in Abu Dhabi emirate', http://www.ead.ae/Tacsoft/FileManager/Misc/2-Water Resources in Abu Dhabi Emirate-EAD.pdf, accessed in December 2011.
27. Raouf, *Water*, pp. 2–3. OECD data for 2002 from World Bank.
28. EAD, 'Water resources'.
29. Raouf, *Water*, p. 4.
30. Woertz, 'Import Dependence', p. 49; *Arabian Business*, 7 November 2009.
31. See e.g. CNA, *Threat*, p. 30; MoE-UAE, *Second National Communications*, p. 30.
32. EIU, *GCC 2020*, p. 16.
33. Woertz, 'Import Dependence', p. 44.
34. IPCC, *Synthesis Report*, p. 45.
35. Kingdom of Bahrain, *Bahrain's Initial National Communication to the UNFCCC. Volume I: Main Summary Report*, GCPMREW (2005), p. 18.
36. Possible accelerated ice cap melting accounted. MoE-UAE, *Second National Communications*, p. 27.
37. IPCC, *Synthesis Report*, p. 13.
38. *Arabian Business*, 11 June 2007; *International Herald Tribune*, 11 June 2007.
39. Brown and Crawford, *Rising*, pp. 10–18.
40. Raouf, *Water*, p. 3.
41. E.g. deportations of tens or hundreds of workers: *Gulf News*, 3 May 2009.
42. Nicholas Stern, *The Stern Review: The Economics of Climate Change* (Cambridge University Press, 2007).
43. Jon Barnett and Suraje Dessai, 'Articles 4.8 and 4.9 of the UNFCCC: Adverse Effects and the Impacts of Response Measures', *Climate Policy*, 2 (2002), p. 234. Decrease in the global demand for fossil fuels can be motivated by binding caps and global emissions trading, taxes and/or a sudden unfolding of an extreme climate change scenario in the future.
44. Shokri Ghanem *et al.*, 'The Impact of Emissions Trading on OPEC', *OPEC Review*, 23 (1999), pp. 104–7. Similarly, the numerical data of a modelling study by Kassler and Paterson from 1997 (*Energy Exporters*), which reaches similar conclusions, are out of date.
45. Tobias A. Persson *et al.*, 'Major Oil Exporters May Profit Rather than Lose in a Carbon-Constrained World', *Energy Policy*, 32 (2007), pp. 6346–7, 6352.
46. Jon Barnett *et al.*, 'Will OPEC Lose from the Kyoto Protocol?', *Energy Policy*, 32 (2004), pp. 2084–7.
47. *New York Times*, 13 October 2009.
48. Barnett *et al.*, 'OPEC Lose?', p. 2086.
49. See: Dennis Kumetat, 'Climate Change in the Persian Gulf—Regional Security, Sustainability Strategies and Research Needs', Conference on Climate Change, Social Stress and Violent Conflict, Hamburg, 19–20 November 2009, pp. 1, 5.
50. Mohamed A. Raouf, *Climate Change Threats, Opportunities, and the GCC Countries*, Middle East Institute Policy Brief, No. 12 (2008), p. 5.

51. Halvald Buhaug *et al.*, *Implications of Climate Change for Armed Conflict*. Social Dimensions of Climate Change (Washington: World Bank, 2008), p. 41.

52. WRI, *CAIT 8.0*.

53. Data for other greenhouse gases n/a. Ibid. Data exclude land use change and international bunkers.

54. Ibid. The US EIA's energy emission projections (low and high) for 2005–2030, estimate a growth of 1.9–2.4% in the Middle East and a global CO_2 emission growth of 1.3–2.1%. Ibid.

55. See e.g. *Guardian*, 29 January 2009; *New York Times*, 27 October 2010.

56. WWF, *Living Planet Report 2012* (2012), pp. 43–4.

57. Gulf Research Center, *Green Gulf Report* (2006), pp. 5–6.

58. Beeah, *Overview of the State of Environment in the Emirate of Sharjah, U.A.E.* (undated), p. 1.

59. Chatham House, *OPEC and Climate Change: Challenges and Opportunities*, Briefing Paper (2005), pp. 40–1.

60. Giddens, *Climate Change*, p. 88.

61. Joergen Fenhann, 'CDM pipeline', UNEP Risoe Centre, http://cdmpipeline.org/, updated on 1 October 2011.

62. Raouf, *Threats*, p. 5.

63. E.g. *Associated Press*, 15 January 2007; interviews in Dubai, October 2008.

64. See e.g. *New Civil Engineer*, 13 February 2008.

65. *The National*, 20 April 2011.

66. Yale University, 'Environmental Performance Index 2010', http://epi.yale.edu/, accessed in December 2010.

67. *The National*, 3 July 2008; 19 August 2010.

68. EIU, *Democracy in Retreat*; Freedom House, 'Freedom in the World'.

69. Steven Wright, *Generational Change and Elite-Driven Reforms in the Kingdom of Bahrain*, Durham Middle East Papers, No. 7 (2006), pp. 11, 25. Nonneman, *Political Reform*, pp. 31, 37.

70. EAD press release, 19 September 2011.

71. Correspondence with report author Justin Gengler, October 2011.

72. Personal communications, November 2010.

73. *The Peninsula*, 5 March 2011.

74. Observation and conversations in the UAE and Qatar, 2008–2012.

75. *Arab News*, 24 February 2010.

76. Interviews in Abu Dhabi, 2009–2010.

77. Al Jazeera English Online, 25 June 2011.

78. IEA, *Oil Information 2010* (Paris: OECD/IEA, 2010); Butt, 'Oil and Gas', p. 237; EIU, *Oman: Country Profile 2008* (2008), p. 29.

79. *Arabian Gazette*, 19 July 2011; *Saudi Gazette*, 9 October 2011; *V&E Climate Change Report*, 7 June 2011.

4. ABU DHABI'S NATURAL SUSTAINABILITY COMPLEX

1. Fatima Al Shamsi, 'Industrial Strategies and Change in the UAE during the 1980s', in Abbas Abdelkarim (ed.), *Change and Development in the Gulf* (Basingstoke and London: Macmillan, 1999), pp. 79, 100.
2. World Bank, 'Gross domestic product 2010', http://www.worldbank.org/; Central Intelligence Agency, 'The World Factbook', https://www.cia.gov/, accessed in October 2011.
3. Average for 2003–2007. IMF, *UAE Appendix 2009*.
4. Data for 2009. US EIA, *UAE*.
5. Butt, 'Oil and Gas', p. 237.
6. IMF, *UAE Appendix 2011*.
7. SCAD, *Statistical Yearbook 2011*, p. 19.
8. EIU, *United Arab Emirates: Country Report* (April 2009), p. 6.
9. IMF cited in *The National*, 13 July 2008; IMF, *UAE Appendix 2009*. Data from ADNOC.
10. Abu Dhabi crude and product exports + UAE's LNG and NGL exports. IMF, *UAE Appendix 2011*.
11. Christopher Davidson, 'The Emirates of Abu Dhabi and Dubai: Contrasting Roles in the International System', *Asian Affairs*, 38 (2007), p. 35; Christopher Davidson, *Dubai: The Vulnerability of Success* (London: Hurst, 2008), p. 79.
12. Sovereign Wealth Fund Institute, 'Sovereign Wealth Fund Rankings'. Estimates vary greatly owing to opacity.
13. *The National*, 13 July 2008.
14. Interview with UAE-based investor, Abu Dhabi, October 2008.
15. Christopher Davidson, *Abu Dhabi: Oil and Beyond* (London: Hurst, 2009), pp. 77–8 and chapter 4.
16. Interview with UAE-based investor, October 2008.
17. Davidson, 'Contrasting', p. 41.
18. Juha Wilén, *Arabiemiraatit: Abu Dhabin rakentaminen*, Helsinki: Finpro toimialakatsaus (2007).
19. Davidson, *Beyond*, p. 86.
20. Abu Dhabi Urban Planning Council, *Plan Abu Dhabi 2030: Urban Structure Framework Plan* (2007), p. 45.
21. *The National*, 14 July 2008.
22. UNFPA, *State of the World Population 2011* (2011); *Gulf News*, 6 October 2009; *UAE Interact*, 19 May 2009.
23. Ministry of Economy of the UAE, 'UAE in Numbers 2007', http://www.economy.gov.ae/, accessed in December 2010; IMF, *UAE Appendix 2009*; SCAD, *Statistical Yearbook 2011*, pp. 118–26; Abu Dhabi Urban Planning Council, *Plan Abu Dhabi 2030: Urban Structure Framework Plan* (2007), p. 45.
24. SCAD, *Statistical Yearbook 2010*, pp. 103, 193.
25. *The National*, 21 June 2009.

26. *The National*, 13 July 2009. Reliable statistics on female unemployment n/a.
27. National Media Council, *UAE Yearbook 2010* (2010), p. 156; *Gulf News*, 4 July 2008.
28. E.g. Prime Minister of the UAE, 'Prime Minister's First e-Session with the Public', http://www.uaepm.ae/, updated in June 2009.
29. *Gulf News*, 4 July 2008.
30. Abu Dhabi Executive Council, *Policy Agenda 2007–2008: The Emirate of Abu Dhabi* (2007), p. 38.
31. See e.g. Frauke Heard-Bey, 'The United Arab Emirates: Statehood and Nation-Building in a Traditional Society', *Middle East Journal*, 59 (2005), p. 375.
32. *The Stream*, 10 October 2011, Al Jazeera English website: http://stream.aljazeera.com/.
33. EIU, *Democracy in Retreat*.
34. Heard-Bey, 'Statehood', p. 358.
35. Davidson, 'Contrasting', p. 37; Gavin Brown (ed.), *OPEC and the World Energy Market: A Comprehensive Reference Guide* (Harlow: Longman, 1991), p. 360.
36. Press reports. Dubai's contribution was 3%. *The National*, 16 June 2010.
37. Davidson, 'Contrasting', pp. 37–8; Davidson, 'Zayed', pp. 43–4; IMF, *UAE Appendix 2011*, p. 56.
38. See e.g.: Davidson, 'Contrasting', p. 43; Neil Patrick, *Nationalism in the Gulf States*, LSE Kuwait Programme Research Paper, No. 5 (2009), p. 18.
39. Davidson, 'Contrasting', p. 43.
40. EIU, *UAE: Country Profile*, p. 5.
41. Butt, 'Oil and Gas', p. 248; US EIA, *UAE* (2007). Note: 2009 version of US EIA's UAE brief is used elsewhere.
42. *Reuters*, 22 September 2010; National Media Council, *UAE Yearbook 2010*, pp. 88, 94.
43. arabianoilandgas.com, 25 March 2009.
44. Fitch and Citigroup cited in: *The National*, 22 November 2008.
45. Japan External Trade Organization press releases, 2 July 2009; 2 August 2010; OPEC, *Annual Statistical Bulletin 2009* (2010).
46. Butt, 'Oil and Gas', pp. 233–4; OPEC, *Annual Statistical Bulletin 2008* (2009).
47. Varies depending on source. Low: US EIA, *UAE*. High: SCAD, *Statistical Yearbook 2011*, p. 71.
48. IEA, *Betwixt Petro-Dollars and Subsidies: Surging Energy Consumption in the Middle East and North Africa States*, IEA Information Paper (2008), p. 5; OPEC, *Annual Statistical Bulletin 2010–11* (2011).
49. US EIA, *UAE*.
50. Phone interview with Hamad Ali Al Kaabi, UAE Representative to the IAEA, November 2010.
51. Butt, 'Oil and Gas', p. 231; *MEED*, 5 January 2007.
52. Abdul Al Kindy, quoted in *The National*, 4 November 2008.
53. IEA, *Energy Balances of Non-OECD Countries* (2010); BP, *Statistical Review of World Energy* (2010).

54. BP, *Statistical Review of World Energy* (2011); OPEC, *Annual Statistical Bulletin 2010–11*; K. Miller, 'ADWEC Winter 2009/2010 Demand Forecast', presentation in Abu Dhabi, 30–31 March 2010.
55. IEA, *Non-OECD Energy Balances* (2010).
56. *The National*, 17 April 2011.
57. Government of the UAE, *Nuclear Energy Policy*.
58. Keith Miller, 'ADWEC Winter 2009/2010 Demand Forecast', presentation in Abu Dhabi, 30–31 March 2010.
59. ENEC website, http://www.enec.gov.ae/, accessed in February 2012; *The National*, 28 December 2009.
60. *The National*, 10 July 2010.
61. See e.g. *Gulf News*, 14 November 2010.
62. *Albawaba*, 18 April 2011; *Zawya*, 26 September 2011; *The National*, 10 January 2012; interview with Robin Mills, Abu Dhabi, December 2011.
63. Jim Krane in *The National*, 19 January 2012.
64. WRI, *CAIT 8.0*. MoE-UAE, *Statistical Report 2003–2007* (2008), in Arabic; SCAD, *Statistical Yearbook 2011*, p. 81.
65. Regulation and Supervision Bureau, *Electricity Tariffs for Large Users in the Emirate of Abu Dhabi*, Information Tariffs (November 2009), p. 3; Regulation and Supervision Bureau, *Annual Report 2010. For the Water, Wastewater and Electricity Sector in the Emirate of Abu Dhabi* (2011), p. 7.
66. IMF, *UAE Appendix 2011*.
67. IMF, *UAE Appendix 2009*.
68. *Reuters*, 26 October 2011.
69. Ibid.
70. *The National*, 10 January 2010.
71. *Kipp Report*, 15 October 2011.
72. Regulation and Supervision Bureau, *Water and Electricity Sector Overview 2008/2009* (2009), p. 4.
73. *Gulf News*, 15 June 2009; SCAD, *Statistical Yearbook 2011*, p. 82.
74. Year not specified. EAD, *Abu Dhabi Water Resources Master Plan* (2009), p. 19.
75. Mahmoud A. Al-Iriani, 'Climate-Related Electricity Demand-Side Management in Oil-Exporting Countries—The Case of the United Arab Emirates', *Energy Policy*, 33 (2004), p. 2359.
76. National Media Council, *UAE Yearbook 2009* (2009), pp. 129–30.
77. *Bloomberg*, 14 February 2011; *Financial Times*, 28 April 2010.
78. *The Citizen*, 26 January 2012; *The National*, 7 October 2011.
79. Keith Miller, *Economic Development and the Growth in Electricity Demand*, presentation in MEED Abu Dhabi Conference, November 2011.
80. Abu Dhabi Executive Affairs Authority website: http://eaa.abudhabi.ae/, accessed in December 2009.
81. *The National*, 22 June 2009; 10 July 2009.
82. *UAE Interact*, 22 March 2009; *The National*, 21 June 2009.

83. *Financial Times*, 21 January 2009.

84. Justin Dargin, *Addressing the Natural Gas Crisis: Strategies for a Rational Energy Policy*, Belfer Center Dubai Initiative Policy Brief (2010), p. 3.

85. EIU, *UAE: Country Profile*, p. 16; *MEED*, 11 January 2008.

86. *The National*, 6 January 2010; UK Trade & Investment, *Power & Water: Dubai and the Northern Emirates, United Arab Emirates (UAE)*, Sector Report (2009), pp. 3–4.

87. *The National*, 15 October 2008; 6 January 2010.

88. *MEED*, 20 March 2008.

89. US EIA, *UAE*; *Gulf Times*, 7 December 2010.

90. US EIA, *UAE*; EIU, *UAE: Country Profile*, p. 17; *The National*, 15 February 2009.

91. *Gulf News*, 23 March 2009; *MEED*, 14 November 2007; *The National*, 17 July 2008; 19 May 2009.

92. Conversation with Sheikh Abdul Aziz bin Ali Al Nuaimi, Helsinki, August 2009.

93. *AMEinfo*, 8 April 2009; *The National*, 4 March 2009.

94. Interview with Abu Dhabi-based journalist, Abu Dhabi, October 2008.

95. *Gulf News*, 1 October 2007.

96. *MEED*, 14 November 2007; supplement 2009.

97. See e.g. *Bloomberg*, 22 September 2011.

98. *Emirates 24/7*, 2 March 2011.

99. *Khaleej Times*, 15 August 2011.

100. Robin Mills in *The National*, 27 March 2012.

101. *The National*, 15 August 2011.

102. Hakim Darbouche and Bassam Fattouh, *The Implications of the Arab Uprisings for Oil and Gas Markets*, Oxford Institute for Energy Studies, MEP 2 (2011), pp. 18, 36.

103. Emirates Center for Strategic Studies and Research, *With United Strength: H.H. Shaikh Zayid Bin Sultan Al Nahyan, The Leader and the Nation* (Abu Dhabi: ECSSR, 2004), p. 62; Michael Herb, *All in the Family: Absolutism, Revolution, and Democracy in the Middle Eastern Monarchies* (Albany: State University of New York, 1999), p. 140.

104. Davidson, 'Zayed', p. 43–4, 52.

105. Bernard Reich (ed.), *Political Leaders of the Contemporary Middle East and North Africa: A Bibliographical Dictionary* (Westport, Connecticut: Greenwood Press, 1990), pp. 516–18; Joseph A. Kéchichian, *Power and Succession in Arab Monarchies: A Reference Guide* (London: Lynne Rienner, 2008), p. 281.

106. Interview with Masdar Director, Abu Dhabi, October 2010.

107. Reich, *Leaders*, p. 518; UNEP, 'Champions of the Earth: 2005 Laureates', http://www.unep.org/champions/laureates/2005/, accessed in December 2010.

108. Pernilla Ouis, 'Greening in the Emirates: The Modern Construction of Nature in the United Arab Emirates', *Cultural Geographies*, 9 (2009), p. 339.

109. Simon Aspinall, 'Environmental Development and Protection in the UAE', in I. Al Abed and P. Hellyer (eds), *United Arab Emirates: A New* Perspective (London: Trident Press, 2001), p. 295; Kéchichian, *Power*, pp. 295, 342.
110. Kéchichian, *Power*, p. 296.
111. Including Davidson, 'Zayed', pp. 48–9; UAE Interact, 'Government'.
112. Davidson, 'Zayed', p. 48; EIU, *United Arab Emirates: Country Report* (June 2009), p. 4.
113. Kéchichian, *Power*, pp. 300–5.
114. Abu Dhabi Executive Affairs Authority website: http://eaa.abudhabi.ae/, accessed in February 2012.
115. National Media Council, *UAE Yearbook 2009*, pp. 27–8.
116. UAE Interact, 'Government'.
117. National Media Council, *UAE Yearbook 2009*, pp. 27–8; EIU, *UAE Country report* (April 2009), p. 26; James O'Brien *et al.*, 'Towards a New Paradigm in Environmental Policy Development in High-Income Developing Countries: The Case of Abu Dhabi, United Arab Emirates', *Progress in Planning*, 68 (2007), p. 209.
118. National Media Council, *UAE Yearbook 2009*, pp. 28–30.
119. Ibid., p. 29; Abu Dhabi Executive Affairs Authority website: http://eaa.abudhabi.ae/; Abu Dhabi Executive Council website: http://gsec.abudhabi.ae/, accessed in December 2010.
120. Interviews with UAE-based journalist, Abu Dhabi, October 2008; and Christopher Davidson, Durham, June 2009.
121. Davidson, 'Contrasting', p. 37; *Survival*, p. 205.
122. EIU, *UAE: Country Profile*, p. 5; National Media Council, *UAE Yearbook 2009*, p. 35; Valérie Marcel, *Oil Titans: National Oil Companies in the Middle East* (Baltimore: Brookings, 2006), p. 78.
123. Butt, 'Oil and Gas', p. 236.
124. Cordesman, *Energy Development*, p. 202.
125. Government of Abu Dhabi, *Economic Vision 2030*.
126. Interview with UAE climate change policy expert, Abu Dhabi, October 2010.
127. O'Brien *et al.*, 'Paradigm', p. 242.
128. Interview with UAE climate change policy expert, Abu Dhabi, October 2010 and personal observations.
129. Interview with Masdar manager, Abu Dhabi, October 2010, among other sources.
130. *Gulf States Newsletter*, 'MBR means business in cabinet reshuffle', Issue 824 (February 2008).
131. Aspinall, 'Environmental', pp. 295–6, 298; ECSSR, 'UAE Experience in Combating Desertification: Strategies and Policies' (2004), http://www.ecssr.ac.ae/, accessed in July 2009; Ministry of Economy of the UAE and UNDP, *Millennium Development Goals. United Arab Emirates*, Second Report (2007), p. 25 (quote).
132. Frederic Launay, *Environmental Situational Assessment for the GCC Countries*,

GRC Research Paper (2006), pp. 70–1; Aspinall, 'Environmental', pp. 297–8; O'Brien *et al.*, 'Paradigm', p. 241; Mohamed A. Raouf, *Economic Instruments as an Environmental Policy Tool: The Case of the GCC Countries*, GRC Gulf Paper (2007), p. 93.

133. Gulf Research Center, *Gulf Yearbook 2006–2007* (2007), p. 470.

134. *The National*, 1 October 2009.

135. Phone interviews with Dubai-based environmental expert, March 2009; UAE-based climate policy expert, June 2009; interview with Masdar expert, Abu Dhabi, October 2010.

136. Law 16/2005. *The National*, 7 November 2011.

137. O'Brien *et al.*, 'Paradigm', p. 241.

138. Abu Dhabi Government website: http://www.abudhabi.ae/, accessed in December 2010.

139. Personal observations and e.g. EAD website: http://www.ead.ae/, accessed in December 2010.

140. O'Brien *et al.*, 'Paradigm', p. 242.

141. Tolba and Saab (eds), *Arab Environment*, p. XXI; EAD, *Annual Report 2009–2010* (2010), p. 14.

142. EAD, *Annual Report 2009–2010*, p. 16.

143. Interviews with UAE climate change experts, Abu Dhabi, October 2010.

144. Phone interview with UAE-based climate expert, June 2009; interviews with UAE climate policy experts, Abu Dhabi, October 2010; CDM-DNA UAE website: http//www.cdm-uae.ae/, accessed in December 2010; EAD press release, 13 January 2010.

145. Personal observation based on interviews in Abu Dhabi and Dubai in 2008–2010.

146. Interview with UAE climate policy expert, Dubai, October 2008.

147. Raouf, *Instruments*, pp. 21, 94, 133–4.

148. Aspinall, 'Environmental', p. 294; O'Brien *et al.*, 'Paradigm', p. 208.

149. Launay, *Assessment*, pp. 70–1.

150. Hamid Rezai *et al.*, 'Coral Reef Status in the ROPME Sea Area: Arabian/Persian Gulf, Gulf of Oman and Arabian Sea', in Clive Wilkinson (ed.), *Status of Coral Reefs of the World: 2004*. Vol. 1 (Australian Institute of Marine Sciences, 2004), p. 164; *Gulf News*, 30 October 2005.

151. O'Brien *et al.*, 'Paradigm', p. 210.

152. Aspinall, 'Environmental', pp. 278, 292.

153. O'Brien *et al.*, 'Paradigm', pp. 233, 243.

154. Interview with Lubna Al Ameri, Supreme Petroleum Council, Abu Dhabi, October 2009.

155. Interviews with UAE climate change experts, Abu Dhabi, October 2009; December 2011.

156. EAD, *Policies and Regulations of Abu Dhabi Emirate, United Arab Emirates* (2008), p. 37.

157. Interview with Masdar director, Abu Dhabi, October 2010.
158. Interview with UAE climate policy expert, October 2008; phone interviews with Dubai-based journalist and environmental expert, March 2009; O'Brien *et al.*, 'Paradigm', p. 210.
159. UNCCD, 'Action Programmes', http://www.unccd.int/actionprogrammes/asia/asia.php, updated on 1 April 2010; EAD State of the Environment website: http://www.soe.ae/, accessed in August 2009.
160. Updated in 2009 for 2009–2013. EAD, *Entity Strategic Plan 2009–2013* (2009).
161. Abu Dhabi Global Environmental Data Initiative website: http://www.agedi.ae/, accessed in August 2009.
162. Interview with UAE climate change policy expert, Abu Dhabi, October 2010; EAD, *Eye on Earth Summit: Abu Dhabi Fact Sheet* (2011), p. 2; EAD, *Environment Vision 2030*, brochure (2012).
163. Abu Dhabi Urban Planning Council, *Plan Abu Dhabi 2030*, pp. 6, 19–20.
164. Interview with UAE climate change policy expert, October 2009.
165. Phone interview with UPC expert, November 2009.
166. Abu Dhabi Urban Planning Council website: http://upc.gov.ae/, accessed in December 2010.
167. For the design phase. *The National*, 29 April 2012.
168. Phone interview with UPC expert, November 2009.
169. Interview with Masdar director, Abu Dhabi, October 2010.
170. Tolba and Saab (eds), *Arab Environment*, p. xxi.
171. Personal observations in UNFCCC conferences in 2008–09 and from local press in 2010.
172. Interview with Masdar director, Abu Dhabi, October 2010.
173. EAD State of the Environment website: http://www.soe.ae/, accessed in August 2009; *The National*, 7 December 2010; correspondence with observer at the Cancún climate conference, December 2010.
174. Interviews with Masdar expert; UAE climate change expert, Abu Dhabi, October 2010.
175. Interviews with UAE-based journalist, October 2009; energy sector expert, December 2011, Abu Dhabi.
176. The 2006 index was a pilot. SEDAC, 'Environmental Sustainability', http://sedac.ciesin.columbia.edu/es/, accessed in December 2010.
177. WWF, *Living Planet Report 2012* (2012), p. 43.
178. Ouis, 'Greening', pp. 337–8, 343.
179. E-mail interview with Habiba Al Marashi, Emirates Environment Group, October 2010.
180. *Gulf News*, 19 March 2008.
181. See: http://www.heroesoftheuae.ae/.
182. Interview, Abu Dhabi October 2010.
183. *Arabian Business*, 14 May 2011.
184. Established in 1999 by the ruler of Dubai. Zayed International Prize for the

Environment website: http://www.zayedprize.org.ae/, accessed in December 2010.

185. Zayed Future Energy Prize website: http://www.zayedfutureenergyprize.com/, accessed in December 2010.

186. 48.9% to 55.3% and 43.8% to 44.4%. EAD, *Annual Report 2009–2010*, pp. 72–3.

187. Personal conversations and observations, Abu Dhabi, February 2010.

188. Ouis, 'Greening', p. 342.

189. Ecoventures, *The State of Environmental Initiatives Among UAE Companies* (2009), p. 3.

190. Elisa Taelman, 'Saadiyat Island Tourist Development Project: Dredging in an Ecologically Sensitive Area', *Terra et Aqua*, issue 116 (2009), pp. 3–11; ALDAR website: http://www.aldar.com/, accessed in December 2010.

191. Ouis, 'Greening', pp. 341–3.

192. The Emirates Environment Group and Emirates Green Building Council in Dubai are also active and visible.

5. ABU DHABI'S CLIMATE CHANGE AND SUSTAINABILITY RESPONSES

1. Bahrain was not included. Susmita Dasgupta *et al.*, *The Impact of Sea Level Rise on Developing Countries: A Comparative Analysis*, World Bank Policy Research Working Paper 4136 (2007), pp. 18–20.

2. MoE-UAE, *Second National Communications*, pp. 25–9, xiii.

3. The EAD and the Stockholm Environment Institute participated actively in both communications.

4. MoE-UAE, *Initial National Communication*, p. 11.

5. MoE-UAE, *Second National Communications*, pp. 25–9, xiii.

6. MoE-UAE, *Initial National Communication*, p. 11.

7. The 9 m scenario accounts for possible accelerated ice cap melting. EAD, *Climate Change: Executive Summary* (2009), p. 43.

8. EAD, *Climate Change: Impacts, Vulnerability & Adaptation* (2009), pp. 98–9, 117.

9. MoE-UAE, *First National Communication*, pp. 47–8.

10. MoE-UAE, *Second National Communications*, pp. 34–5.

11. FAO, *Aquastat online*, http://www.fao.org/nr/water/aquastat/dbase/index.stm, accessed in December 2010.

12. Musa N. Nimah, 'Water Resources' in Tolba and Saab (eds): *Arab Environment*, p. 65; Brown and Crawford, *Rising*, p. 11; FAO, *Aquastat online*, accessed in February 2012.

13. WAM, 22 March 2009.

14. MoE-UAE, *First National Communication*, p. 39; EAD, *Climate Change: Summary*, p. 7.

15. National Media Council, *UAE Yearbook 2010*, p. 57.

16. *The National*, 7 May 2009; 21 June 2009; 22 August 2011.

17. EAD, 'Water resources'; Mohamed Dawoud in *The National*, 22 March 2012.

18. Five large plants in Abu Dhabi and one in Fujairah. Interview with Mohamed Dawoud, EAD, Abu Dhabi, October 2010. Reserve data for 2007. EAD, 'Water resources'.

19. *Reuters*, 20 December 2010; interview with Mohamed Dawoud, Abu Dhabi, December 2011.

20. *Bloomberg*, 16 July 2012.

21. Renee Richer, *Conservation in Qatar: Impacts of Increasing Industrialization*, CIRS Occasional Paper (2008), p. 6; EIU, *GCC 2020*, p. 14.

22. EAD, *Climate Change: Summary*, p. 7.

23. *The National*, 10 November 2010.

24. Interview with Mohamed Dawoud, Abu Dhabi, December 2011.

25. Jean G. Chatila, 'Municipal and Industrial Water Management', in Hussein Abaza *et al.* (eds), *Arab Environment: Green Economy. Sustainable Transition in a Changing Arab World*, Arab Forum for Environment and Development (2011), p. 75.

26. EAD, *Water Resources Master Plan*, pp. 18–24.

27. MoE-UAE, *First National Communication*, pp. 39, 42.

28. Adil A. Bushnak, 'Desalination', in Mohamed el-Ashry *et al.* (eds), *Arab Environment: Water. Sustainable Management of a Scarce Resource*, Arab Forum for Environment and Development (2010), p. 133; interview with Mohamed Dawoud, Abu Dhabi, December 2011.

29. Interview with Mohamed Dawoud, Abu Dhabi, October 2010.

30. *The National*, 19 November 2011.

31. EIU, *GCC 2020*, p. 16; *Gulf News*, 20 September 2011; *The National*, 10 November 2010.

32. Interview with Mohamed Dawoud, Abu Dhabi, December 2011.

33. *Khaleej Times*, 10 November 2011.

34. Eckart Woertz, 'Food Inflation in the GCC Countries', *Gulf Monitor*, 2 (2008), p. 16; *Gulf News*, 9 July 2008; *Reuters*, 1 July 2008.

35. Joachim Braun and Ruth Meinzen-Dick, *'Land grabbing' by Foreign Investors in Developing Countries: Risks and Opportunities*, IFPRI Policy Brief, No. 13 (2009).

36. E.g. *Gulf News*, 9 March 2011.

37. *Reuters*, 1 September 2009.

38. *The National*, 18 May 2010.

39. *Financial Times*, 16 November 2010; *The National*, 1 March 2010.

40. MoE-UAE, *Second National Communication*, p. 29.

41. EAD, *Climate Change: Vulnerability*, p. 57.

42. Communication from EWS/WWF, 16 January 2011.

43. ADNOC, *Developing our Natural Resources Responsibility: Abu Dhabi National Oil Company (ADNOC) 2010 Sustainability Report* (2011), p. 28.

44. Other: 1%. EWS/WWF *et al.*, *UAE Ecological Footprint Initiative* (2010), pp. 4–5.

45. Interview with UAE climate change expert, Abu Dhabi, October 2010.

46. ADNOC, *Health, Safety & Environment Report 2008* (2009), p. 16; ADNOC website: http://www.adnoc.ae/, accessed in December 2010.

47. Ayoub M. Kazim, 'Assessments of Primary Energy Consumption and its Environmental Consequences in the United Arab Emirates', *Renewable and Sustainable Energy Reviews*, 11 (2007), p. 440.

48. Haris Doukas *et al.*, 'Renewable Energy Sources and Rationale Use of Energy Development in the Countries of GCC: Myth or Reality?', *Renewable Energy* 31 (2006), p. 766. Kazim, 'Energy', pp. 440–1.

49. OPEC, Statement to the United Nations Climate Change Conference (COP13), Bali, 14 December 2007.

50. Joergen Fenhann, 'CDM pipeline', updated on 1 September 2012. Price data: *Reuters*, 23 November 2011. Emission data for 2007: WRI, *CAIT 8.0.*

51. MoE-UAE, *First National Communication*, chapters 4 and 5.

52. Interview with UAE climate expert, Abu Dhabi, October 2010.

53. See e.g.: Waleed El Malik quoted in *The National*, 17 May 2008.

54. Interview with Masdar director, October 2010. Names added by the author.

55. Interview with UAE climate policy export, Abu Dhabi, October 2009.

56. Interview with UAE climate policy expert, Abu Dhabi, October 2010.

57. EAD, *Annual Report 2009–2010*, p. 36.

58. Interview with UAE climate change expert, Abu Dhabi October 2010.

59. Interviews with UAE climate change expert and Masdar expert, Abu Dhabi, October 2010.

60. Seminar *Beyond Copenhagen: Environmental Diplomacy at Crossroads*, Zayed University, Abu Dhabi, 4 May 2010. Chatham House rule applied.

61. National Media Council, *UAE Yearbook 2009*, pp. 122–3.

62. MoE-UAE, *First National Communication*, p. 1.

63. Internal memo of the Finnish Ministry of Foreign Affairs, 4 February 2010.

64. *Khaleej Times*, 15 January 2012.

65. Interviews with Saad Al Numairy, Ministry of Environment and Water, Dubai, October 2009; UAE climate policy expert, Abu Dhabi, October 2010.

66. Thani Al Zayoudi, 'Key UAE Mitigation Initiatives', mitigation side event, Bonn climate talks, 1 August 2010.

67. *Gulf Today*, 27 September 2011.

68. EWS/WWF *et al.*, *UAE Ecological Footprint*, pp. 4–5.

69. Interview with UAE climate change expert, October 2010.

70. EWS/WWF *et al.*, *UAE Ecological Footprint*, pp. 6–7.

71. Large parts of the case study have been published in Mari Luomi, 'Abu Dhabi's Alternative-Energy Initiatives: Seizing Climate-Change Opportunities', *Middle East Policy*, 16 (2009), pp. 105–9.

72. Zayed Future Energy Prize website: http://www.zayedfutureenergyprize.com/, accessed in February 2011.

73. Sultan al-Jaber in: EWS/WWF, *Dar al Khair*, 24 (2010), p. 6.

74. Interview with Masdar expert, October 2010.

75. Sam Nader, 'Paths to a Low-Carbon Economy—the Masdar Example', *Energy Procedia*, 1 (2009), p. 3952; Masdar website: http://www.masdar.ae/, accessed in July 2009.
76. *Gulf News*, 28 October 2007.
77. Masdar press release, 9 February 2008.
78. *Arabian Business*, 9 June 2009.
79. Nader, 'Masdar', pp. 3953–4; Ryan Tompkins, presentation at Arabian Peninsula business event, Finpro, Helsinki, 29 April 2009.
80. *Masdar*, 9 February 2008.
81. Nader, 'Masdar', pp. 3952–3.
82. Masdar press release, 25 October 2011.
83. *The National*, 21 November 2011.
84. General Electric press release, 20 January 2009; *Reuters*, 22 July 2008.
85. Masdar website: http://www.masdar.ae/, accessed in July 2009; Torresol Energy website: http://www.torresolenergy.com/, accessed in November 2011.
86. Jari Varjotie, presentation at Arabian Peninsula business event, Helsinki: Finpro, 29 April 2009.
87. *The National*, 16 October 2008.
88. Masdar PV press release, 18 March 2009.
89. *The National*, 24 January 2012.
90. *Gulf News*, 1 June 2009.
91. *Wind Power Monthly*, 9 June 2011; *The National*, 24 January 2012.
92. BP press release, 21 January 2008; *Zawya Dow Jones*, 18 January 2011.
93. Juha Wilén, *Arabiemiraatit Abu Dhabin rakentaminen*, Helsinki: Finpro toimialakatsaus (2007), p. 10; Nader, 'Masdar', p. 3955.
94. See e.g.: *Arabian Business*, 22 June 2009; Masdar press release, 20 January 2010; *Reuters*, 17 February 2010; Fenhann, 'CDM Pipeline', updated on 1 September 2012.
95. *Reuters*, 18 November 2008; *The National*, 11 June 2008; 10 December 2008; *UAE Interact*, 29 October 2008.
96. *MEED*, 14 June 2011; *Petroleum Economist*, 4 November 2010; *The National*, 19 January 2010.
97. *Zawya*, 18 January 2012.
98. *Saudi Gazette*, 24 June 2012.
99. Steven Caton and Nader Ardalan, *New Arab Urbanism: The Challenge to Sustainability and Culture in the Gulf*, Harvard University (2010), p. 62.
100. Correspondence with UAE-based ambassador, Abu Dhabi, January 2011.
101. Interview, Abu Dhabi October 2010.
102. *The National*, 12 November 2008.
103. *Utilities-ME*, 8 November 2010.
104. Conversation in Masdar City, Abu Dhabi, October 2009.
105. *The National*, 1 March 2010; 11 January 2011.
106. Correspondence with Craig Field, Qatar Solar Technologies, January 2011.

107. Conversations in Abu Dhabi, May 2010.
108. *Arabian Business*, 28 February 2010.
109. *Bloomberg*, 21 June 2010; *Chronicle of Higher Education*, 2 August 2010; *Greentech Solar*, 7 May 2010.
110. *Arabian Business*, 28 February 2010.
111. *UAE Interact*, 30 June 2010; *Arabian Business*, 26 July 2010.
112. *The National*, 4 January 2010.
113. See e.g. *Utilities-ME*, 8 November 2010.
114. Correspondence with Middle East renewable energy expert, August 2010.
115. *Utilities-ME*, 8 November 2010.
116. Personal observations and conversations, Masdar City, Abu Dhabi, October 2010.
117. Sheikh Abdullah bin Zayed Al Nahyan, statement in joint high-level segment, 16th Conference of the Parties, UNFCCC, Cancún, 9 December 2010.
118. *The National*, 1 December 2010.
119. *Bloomberg*, 16 September 2011.
120. *The National*, 23 November 2011.
121. *The Telegraph*, 10 June 2010.
122. *The National*, 28 December 2010; WAM, 13 January 2012.
123. *The Grid/The National*, 18 February 2010.
124. *UAE Interact*, 24 August 2011.
125. *Arabian Business*, 12 November 2011; *Kipp Report*, 13 November 2011.
126. Interview with Masdar director, Abu Dhabi, October 2010.
127. UNEP website: http://www.unep.org/champions/, accessed in June 2012.
128. Masdar press release, 20 October 2011.
129. Internal memo of the Finnish Ministry of Foreign Affairs, 4 February 2010.
130. Interview with Masdar Institute faculty member, Abu Dhabi, October 2008.
131. Sources include the Masdar Initiative's web pages; Masdar's brochures from 2008–2009; Masdar City presentation, Helsinki, April 2009; ADNOC's CEO in: *Time Magazine*, 13 February 2008; and Sultan al-Jaber in: *The National*, 29 May 2008.
132. Conversations with Masdar Institute staff and students, Abu Dhabi, October 2010; Masdar Institute website, http://www.masdar.ac.ae/, accessed in November 2011.
133. Interview with Masdar director, Abu Dhabi, October 2010.
134. E.g. Al-Jaber, quoted in Mubadala press release, 2 July 2007.
135. Khaled Awad in NPR web article, 5 May 2008.
136. *New York Times*, 25 September 2010.
137. Hertog and Luciani, *Sustainability*, p. 31.
138. *The National*, 27 November 2011.
139. Parts of the case study have been published in Luomi, 'Abu Dhabi's Alternative-Energy', pp. 109–12, and Luomi, 'Economic and Prestige'.
140. Phone interview with Hamad Ali Al Kaabi, UAE permanent representative to the IAEA, November 2010.

141. *EIU ViewsWire*, 28 May 2008; *Khaleej Times*, 21 July 2007; Government of the UAE, *Nuclear Energy Policy*.
142. Government of the UAE, *Nuclear Energy Policy*, pp. 1–2, 9.
143. Phone interview with Hamad Ali Al Kaabi, November 2010.
144. Government of the UAE, *Nuclear Energy Policy*, p. 1.
145. *Emirates Business 24/7*, 23 June 2008.
146. *MEED*, 7 July 2009.
147. *Financial Times*, 27 December 2009.
148. US Embassy in Abu Dhabi, 'UAE Nuclear Tender Update' (19 August 2009), via http://www.cablesearch.org/.
149. Emirates Nuclear Energy Corporation press release, 22 April 2010.
150. *Nuclear Engineering International*, 10 January 2011.
151. *Arabian Business*, 20 July 2012.
152. Interview with Christer Viktorsson, Federal Authority for Nuclear Regulation, Abu Dhabi, May 2010.
153. Emirates Nuclear Energy Corporation press release, 27 December 2009.
154. Giacomo Luciani, 'The Gulf Countries and Nuclear Energy', *Gulf Monitor*, 6 (2007).
155. Working group on *the Nuclear Question in the Middle East: Regional Perspectives*, Center for International and Regional Studies, Georgetown University in Qatar, Doha, 7 November 2010.
156. Government of the UAE, *Nuclear Energy Policy*, p. 1.
157. Seminar on *Enhancing the EU-GCC relations within a New Climate Regime: Prospects and Opportunities for Cooperation*, Gulf Research Centre and Centre for European Policy Studies, Brussels, 26 February 2009.
158. *Gulf News*, 19 October 2008.
159. *The Wall Street Journal*, 4 October 2009.
160. Government of the UAE, *Nuclear Energy Policy*, p. 3; *The National*, 2 August 2009.
161. *Emirates 24/7*, 25 September 2011.
162. Conversation with a high-level US diplomat, Helsinki, September 2008.
163. Government of the UAE, *Nuclear Energy Policy*, pp. 9–10.
164. Christopher M. Blanchard and Paul K. Kerr, *The United Arab Emirates Nuclear Program and Proposed U.S. Nuclear Cooperation*, Congressional Research Service (17 July 2009), p. 4.
165. World Nuclear Association, 'Nuclear Power in the United Arab Emirates, http://www.world-nuclear.org/, updated in July 2009.
166. ECSSR, 'H.E. Ambassador Hamad Ali Al-Kaabi, Profile', http://www.ecssr.ac.ae/, accessed in August 2009.
167. Personal conversations in Abu Dhabi in 2009–2011 and e.g. *Wall Street Journal*, 2 April 2009.
168. World Nuclear Association, 'UAE'.
169. *MEED*, 14 October 2008.

170. Federal Authority for Nuclear Regulation website; http://fanr.gov.ae/, accessed in January 2011; *UAE Interact*, 23 February 2010.

171. *AMEinfo*, 23 February 2011; *Wall Street Journal*, 2 April 2009.

172. IISS, *Shadow*, p. 55.

173. World Nuclear Association, 'UAE'.

174. *The Gulf*, 8 June 2009.

175. Blanchard and Kerr, *Cooperation*, p. 8; *Stratfor*, 15 January 2008; *Los Angeles Times*, 16 December 2008.

176. *New York Times*, 21 May 2009.

177. *Associated Press*, 8 July 2009.

178. IISS, *Shadow*, p. 53; *Khaleej Times*, 15 May 2008; *Stratfor*, 16 June 2009.

179. Emirates Nuclear Energy Corporation press release, 27 December 2009.

180. *The National*, 28 December 2009.

181. *The National*, 19 October 2010.

182. *AMEinfo*, 9 July 2011.

183. Khaldoon al-Mubarak in: Emirates Nuclear Energy Corporation press release, 27 December 2009.

184. Hamad Al Kaabi, *Challenges Faced by Developing Countries in Nuclear Power Deployment: UAE Approach*, presentation in an IAEA meeting, 27 October 2009.

185. Phone interview with Hamad Ali Al Kaabi, November 2010.

186. Blanchard and Kerr, *Cooperation*, p. 10.

187. IISS, *Shadow*, p. 55; *Wall Street Journal*, 2 April 2009.

188. Embassy of the UAE in Washington website: http://www.uae-embassy.org/, accessed in January 2011.

189. Government of the UAE, *Nuclear Energy Policy*, pp. 1, 13.

190. *Gulf News*, 16 January 2009.

191. FANR press release, 7 June 2011; *Khaleej Times*, 16 June 2011.

192. Personal communications, November 2010.

193. ENEC website: http://www.enec.gov.ae/, accessed in November 2011.

194. *Bloomberg*, 16 June 2011.

195. Interview with Christer Viktorsson, Abu Dhabi, May 2010.

196. *Arabian Business*, 20 May 2009.

197. See e.g.: Sharon Squassoni and Kee Hon Chung, 'Nuclear Choices in the Middle East', *Nuclear in Focus blog*, CSIS website: http://www.csis.org/blog/, dated 14 October 2010; *Arabian Business*, 16 June 2010.

6. QATAR'S NATURAL SUSTAINABILITY COMPLEX

1. BP, *Statistical Review of World Energy* (2011); *MEED*, 5 June 2009.

2. *Gulf Times*, 29 November 2011.

3. Abdullah al-Attiyah quoted in *Reuters*, 10 January 2010; *MEED*, 5 June 2009.

4. *Gulf Times*, 27 July 2010.

5. Estimates. Central Intelligence Agency, 'The World Factbook', https://www.cia.

gov/, accessed in October 2011; IMF, 'World Economic Outlook Database', http://www.imf.org/external/pubs/ft/weo/2011/02/weodata/weoselgr.aspx updated in September 2011.

6. Qatar National Bank, *Economic Insight*, p. 4.

7. EIU, *Qatar: Energy Report* (1 December 2009).

8. IMF, *Qatar: Statistical Appendix 2010*. IMF Country Report No. 10/62 (2010).

9. Qatar Central Bank, *The Thirty Fourth Annual Report* (2011), p. 177.

10. EIU, *Qatar: Country Report* (December 2010), p. 5; US EIA, *Qatar* (2009); *MEED*, 5 June 2009.

11. Qatar Central Bank, *Annual Report*, p. 177.

12. Sovereign Wealth Fund Institute, 'Qatar Investment Authority', http://www.swfinstitute.org/, accessed in December 2010.

13. *Reuters*, 21 September 2009, 2 November 2009; *The Peninsula*, 29 August 2009.

14. UNCTAD, *World Investment Report 2010. Investing in a Low-Carbon Economy* (2010), p. 27.

15. *MEED*, 5–11 February 2010.

16. GSDP, *Qatar National Development Strategy 2011–2016* (2011), pp. 52–4, 93, 96; *MEED*, 5–11 February 2010; conversations in Doha, January 2010.

17. See also J.E. Peterson, 'Qatar and the World: Branding for a Micro-State', *Middle East Journal*, 60 (2006); Uzi Rabi, 'Qatar's Relations with Israel: Challenging Arab and Gulf Norms', *Middle East Journal*, 63 (2009), p. 458.

18. Central Intelligence Agency, *The World Factbook*, https://www.cia.gov/, accessed in October 2011; UNCTAD, *World Investment Report 2010*.

19. *Gulf Times*, 1 April 2008.

20. EIU, *Qatar: Country Profile 2009* (2009), p. 22.

21. Qatar National Bank, *Economic Insight*, p. 18.

22. *Arabian Business*, 15 September 2010; United Development Company website: http://www.udcqatar.com/, accessed in January 2011; EIU, *Qatar: Country Profile*, p. 26.

23. *Al Bawaba*, 14 January 2010.

24. Al Jazeera English Online, 22 November 2009; *Arabian Business*, 11 February 2010; *MEED*, 5–11 February 2010; *The Peninsula*, 26 October 2009.

25. Ahmed Elmagarmid, 'Qatar Computing Research Institute', presentation, 29 May 2011.

26. World Bank, *World Development Indicators*, accessed in October 2011; Qatar Statistics Authority, 'Qatar Census 2010', http://www.qsa.gov.qa/, accessed in January 2011; EIU, *Qatar: Country Profile*, p. 11.

27. *Construction Week Online*, 4 December 2010; *MEED*, 5–11 February 2010.

28. Conversations in Doha, January 2011.

29. See e.g. Kamrava, 'Factionalism', pp. 406–7.

30. Qatar Statistics Authority, 'Population and Social Statistics. Labor Force Sample Survey', http://www.qsa.gov.qa/, dated December 2008.

31. Ananthakrishnan Prasad, World Bank, presentation at Georgetown University Qatar, 28 November 2011.

32. Qatar Statistics Authority, *Labor Force Sample Survey Results, 2008: Analytical Summary*, Population & Social Statistics Department (2009 or 2010).

33. EIU, *Qatar: Country Profile*, p. 7; Freedom House, 'Freedom in the World 2010'.

34. Louay Bahry, 'Elections in Qatar: A Window of Democracy Opens in the Gulf', *Middle East Policy*, 6 (1999), pp. 118–9; EIU, *Qatar: Country Profile*, pp. 4–5, 8; *Gulf News*, 1 November 2011.

35. Qatar News Agency, 6 September 2011.

36. Kamrava, 'Factionalism', pp. 404–6, 417; Rabi, 'Relations', p. 444.

37. Kéchichian, *Power*, p. 189; EIU, *Qatar: Country Profile*, p. 17.

38. BP, *Statistical Review of World Energy* (2011).

39. US EIA, *Qatar* (2009).

40. BP, *Statistical Review of World Energy* (2011).

41. Lower estimate: Central Intelligence Agency, Central Intelligence Agency, 'The World Factbook', https://www.cia.gov/, accessed in October 2011. Higher estimate: BP, *Statistical Review of World Energy* (2011).

42. *Reuters*, 20 March 2010; Ananthakrishnan Prasad, World Bank, presentation at Georgetown University Qatar, 28 November 2011.

43. Nasser Al-Othman, *With their Bare Hands: The Story of the Oil Industry in Qatar* (Harlow: Longman, 1984), p. 125.

44. EIU, *Qatar: Country Profile*, pp. 17–18.

45. US EIA, *Qatar* (2009).

46. Justin Dargin, *The Dolphin Project: The Development of a Gulf Gas Initiative*, Oxford Institute for Energy Studies, NG 22 (2008), pp. 2, 13, 16.

47. BP, *Statistical Review of World Energy* (2011).

48. IEA, *Non-OECD Energy Balances* (2011).

49. BP, *Statistical Review of World Energy* (2011).

50. Justin Dargin, 'Qatar's Natural Gas: The Foreign-Policy Driver', *Middle East Policy*, 3 (2007), p. 136.

51. EIU, *Qatar: Country Profile*, pp. 11–12.

52. Dargin, *Dolphin*, pp. 1, 6, 48; *The Pioneer* (November-December 2009), p. 7; US EIA, *Qatar Country Analysis Brief* (January 2011).

53. Dargin, *Dolphin*, p. 14.

54. US EIA, *Qatar* (2011).

55. Ibid.

56. Environmental Capital, Wall Street Journal Blog: http://www.blogs.wsj.com/, dated 28 October 2009; Qatargas website http://www.qatargas.com/, accessed in December 2011.

57. US EIA, *Qatar* (2011); EIU, *Qatar: Energy Report* (25 October 2011).

58. Al-Othman, *Bare Hands*, p. 115.

59. BP, *Statistical Review of World Energy* (2011).

60. IEA, *Non-OECD Energy Balances* (2010).

61. *MEED*, 28 March 2010; Qatar Electricity and Water Company website: http://www.qewc.com/, accessed in January 2011.

62. *MEED*, 5 June 2011.

63. EIU, *Qatar: Energy Report* (2011).

64. *MEED*, 28 March 2010.

65. BP, *Statistical Review of World Energy* (2010).

66. *The Peninsula*, 21 May 2012; 28 May 2012.

67. *The Peninsula*, 4 November 2007.

68. US Embassy in Doha, 'Qatar's Plans to Pursue Nuclear Energy' (17 December 2008), via http://www.cablesearch.org/.

69. *MEED*, 8 February 2010.

70. Oxford Business Group, *The Report: Qatar 2010* (2010), p. 139; Sogreah, *Annual Report 2009* (2010), p. 6; *The Peninsula*, 5 October 2010; *Utilities-me.com*, 9 July 2010.

71. GSDP, *Development Strategy*, p. 85.

72. IEA, *World Energy Outlook 2008* (2008), p. 300.

73. WRI, *CAIT 8.0*.

74. Ali Hamed Al Mulla, *Chapter 4: Climate Change and Human Development in Qatar: Issues, Challenges and Opportunities*, unpublished, drafted for Qatar's *Second Human Development Report* (2009), p. 10.

75. EIU, *Qatar: Energy Report* (2011).

76. GSDP, *Development Strategy*, pp. 83, 218; *Arabian Business*, 5 November 2009.

77. Hassan Ibrahim al-Mohannadi, cited in: *MENAFN*, 19 March 2009. OECD data from World Bank.

78. GSDP, *Development Strategy*, p. 218.

79. Electricity data, December 2009: EIU, *Qatar: Energy Report* (2009). Water data, 2005–2008: GSDP, *Second HDR*, p. 48. Electricity in Finland: Energiamarkkinavirasto website: http://www.sahkonhinta.fi/, accessed in December 2011.

80. Kahramaa, *Statistics Report 2008* (2008), p. 18.

81. Estimate. IMF, *Qatar: Statistical Appendix*.

82. GSDP, *Development Strategy*, pp. 81, 218.

83. Notably, calculated with a total population of 0.8 million. OPEC, *World Oil Outlook 2011* (2011), pp. 80–1.

84. United Nations, *The United Nations Regional Commissions and the Climate Change Challenges* (2009), pp. 65, 69.

85. See e.g. *Gulf Times*, 28 December 2008.

86. Presentation by Kahramaa representative for SustainableQatar, 4 June 2012.

87. Dargin, 'Qatar's Gas', pp. 139–40.

88. Henry hub prices. Dargin, *Dolphin*, p. 9; *MEED*, 11–17 September 2009; US EIA, 'Natural Gas Weekly Update', (1 December 2011); IEA, *Betwixt*, pp. 9–10; Dolphin Energy website: http://www.dolphinenergy.com/, accessed in December 2010.

89. Dargin, *Dolphin*, pp. 9–11, 48; EIU, *Qatar: Country Profile*, p. 19.

90. *Bloomberg*, 30 November 2010; *Gulf Times*, 7 December 2010.

91. See *Bloomberg*, 3 July 2009.

92. El-Katiri, Laura, *Interlinking*, p. 26.

93. *The Peninsula*, 8 July 2009; 6 September 2010.

94. Emile A. Nakhleh, "'The Creation of Qatar" by Rosemarie Said Zahlan', review, *International Journal of Middle East Studies*, 16 (1984), p. 295.

95. Herb, *Family*, p. 109.

96. Jill Crystal, *Oil and Politics in the Gulf: Rulers and Merchants in Kuwait and Qatar* (Cambridge University Press, 1995), pp. 30–1.

97. Kéchichian, *Power*, pp. 202, 208.

98. Carnegie Endowment, 'Qatar', Arab Political Systems, http://carnegieendowment.org/, accessed in January 2011; interview with informed source, Doha, July 2012.

99. Qatar Investment Authority website: http://www.qia.qa/, accessed in April 2010; Amiri Diwan website: http://www.diwan.gov.qa/, accessed in January 2011.

100. Kéchichian, *Power*, pp. 202, 216–17.

101. E.g.: EIU, *Qatar: Country Profile*, p. 5.

102. Conversation in Doha, November 2009.

103. Bahry, 'Elections', p. 122.

104. HH Sheikha Moza Office website: http://www.nozahbintnasser.qa/, accessed in January 2011.

105. Conversations in Doha, November 2009.

106. EIU, *Qatar: Country Profile*, p. 6; Kéchichian, *Power*, p. 200; Ministry of Foreign Affairs of Qatar website: http://english.mofa.gov.qa/, accessed in April 2010; Qatar News Agency, 20 September 2011. In 2012, Abdullah al-Attiyah no longer used the title Deputy Prime Minister.

107. Water and electricity issues were attached to the energy minster's portfolio in 1999. *IHS Global Insight*, 19 January 2011.

108. *Gulf Times*, 4 April 2007; Supreme Education Council of Qatar, 'Interview with SEC Member Mohammed Saleh Al-Sada', http://www.english.education.gov.qa/content/resources/detail/3277, dated 2 April 2006.

109. Kamrava 'Factionalism', pp. 402–3.

110. Ministry of Foreign Affairs of Qatar website: http://english.mofa.gov.qa/, accessed in April 2010.

111. *Gulf Times*, 2 July 2008.

112. EIU, *Qatar: Country Profile*, pp. 5–6; Carnegie Endowment, 'Qatar'; Kamrava, 'Factionalism', p. 417.

113. EIU, *Qatar: Country Report* (January 2010), p. 4; Ministry of Foreign Affairs of Qatar website: http://english.mofa.gov.qa/, accessed in April 2010.

114. Carnegie Endowment, 'Qatar'.

115. Kamrava, 'Factionalism', p. 414; Ministry of Foreign Affairs of Qatar website: http://english.mofa.gov.qa/, accessed in April 2010.

116. Kamrava, 'Factionalism', pp. 414–15.

117. *APS Diplomat Operations in Oil Diplomacy*, 30 October 2000.

118. Hukoomi, Qatar e-Government website: http://www.gov.qa/, accessed in April 2010.
119. *Qatar Tribune*, 9 June 2011; GSDP website: http://www.gsdp.gov.qa/, accessed in December 2011.
120. Interview with Qatar-based PR specialist, Doha, November 2010.
121. Interview with Richard Leete, GSDP, Doha, November 2010.
122. Interview with Qatar-based PR specialist, Doha, November 2010.
123. GSDP, *National Vision*, p. 11 and *passim*.
124. Interview with Elina Lehtinen, Crisis Management Initiative, formerly based in Qatar, Helsinki, June 2010. Prime Minister Hamad bin Jassim is also known to have been closely involved in the process.
125. GSDP, *National Vision*, p. 34; GSDP press release, 5 February 2009.
126. Interviews with Richard Leete and Aziza Al Khalaqi, GSDP, Doha, November 2010.
127. Interviews in Doha in November 2010 with: Aziza Al Khalaqi and Richard Leete, GSDP; Qatar-based renewable energy experts; Yousef Al Horr, BQDRI; Qatar-based renewable energy expert; Katrin Scholz-Barth, SustainableQatar; Sam Pickering, BluuGreen Qatar; and Qatar-based PR specialist.
128. United Nations, *Johannesburg Summit 2002. Qatar Country Profile* (2002), p. 11. Data in the profile: from 1994–1996. Atkins, *Qatalum EIA Report* (December 2006), p. 2–1.
129. GSDP, *Second HDR*, p. 19.
130. UNEP, *Overview of Land-Based Sources and Activities Affecting the Marine Environment in the ROPME Sea Area*, UNEP Regional Seas Reports and Studies No. 168 (1999), p. 50.
131. GSDP, *Second HDR*, p. 19; Embassy of the State of Qatar in Washington website: http://www.qatarembassy.net/, accessed in April 2010.
132. Hukoomi, Qatar e-Government website: http://www.gov.qa/, accessed in April 2010; interview with Aziza Al Khalaqi, November 2010.
133. Richer, *Conservation*, pp. 11–12.
134. GSDP, *Second HDR*, p. 86.
135. Includes institutions under the Ministry. Correspondence with Aziza Al Khalaqi, GSDP, January 2011.
136. Interviews with Qatar-based PR specialist and IGO and GSDP representatives, Doha, November 2010.
137. Qatar Petroleum website: http://www.qp.com.qa/, accessed in January 2011; interview with Qatar Petroleum manager, Doha, October 2009.
138. Interview with Qatar-based PR specialist, November 2010; Atkins, *Qatalum EIA Report*, pp. 2–1; 2–3.
139. Raouf, *Instruments*, p. 89.
140. Ibid., pp. 89–90, 137; Embassy of the State of Qatar in Washington website: http://www.qatarembassy.net/, accessed in April 2010.
141. GSDP, *Second HDR*, p. 19; GSDP, *National Vision*, p. 30.

142. Farid Chaaban, 'Air Quality', in Tolba and Saab (eds), *Arab Environment*, p. 57.

143. Richer, *Conservation*, p. 11; interview with Qatar-based PR specialist, November 2010; conversation with environmental experts, December 2011, Doha.

144. GSDP, *Second HDR*, pp. 19, 21; Richer, *Conservation*, pp. 2, 12.

145. Raouf, *Instruments*, p. 90.

146. UN Division for Sustainable Development, *CSD-14/15 Thematic Profiles: Industry* (2007), p. 1; UN Division for Sustainable Development, *Atmosphere* (2007), p. 2.

147. Environment Minister al-Midhadi in: *RasGas Magazine*, issue 29 (2010), p. 8.

148. Correspondence with Aziza Al Khalaqi, January 2011.

149. Rezai *et al.*, 'Coral Reef Status', p. 164.

150. Al Mulla, *Climate Change*, p. 11.

151. GSDP, *National Vision*, pp. 8, 30–33, quote from p. 8; Sheikh Hamad bin Jabor bin Jassim Al Thani, speech in seminar *Towards Qatar's National Strategy: Issues and Challenges*, Doha, 28 October 2008.

152. GSDP, *Development*, pp. 20, 26, 226.

153. GSDP, *Second HDR*, p. 23.

154. SEDAC, 'Environmental Sustainability', http://sedac.ciesin.columbia.edu/es/, accessed in December 2010.

155. WWF, *Living Planet Report 2010*, p. 36.

156. GSDP, *Second HDR*, *passim*.

157. Based on a newspaper archive search covering *Gulf Times* and *The Peninsula* from 2004 until mid-2010.

158. Interviews with Benno Boer, Doha, November 2010; Richer, *Conservation*, pp. 16, 19.

159. *AMEinfo*, 12 April 2010.

160. Conversation with Friends of the Environment Centre representatives, Doha, November 2011.

161. *Arabian Business*, 28 February 2009; Qatar Green Building Council press release, 6 July 2009.

162. Conversation with Qatar Green Center representative, Doha, February 2012.

163. Interviews with Benno Boer, UNESCO, Doha, November 2010.

164. *Gulf Times*, 14 May 2008.

165. *The Peninsula*, 7 June 2010.

166. Interviews with Western IGO and NGO representatives, Doha, November 2010; conversations, Doha, January 2012.

167. English translation of exact phrase: 'It is important for this person to be concerned for the environment and for protecting it'. Correspondence with World Values Survey Qatar 2010 author, October 2011.

168. Giddens, *Climate Change*, *passim*.

7. QATAR'S CLIMATE CHANGE AND SUSTAINABILITY RESPONSES

1. Dasgupta *et al.*, *Sea Level Rise*, pp. 18–20, 41; Ministry of Environment (MoEnv) of Qatar, *Initial National Communication to the UNFCCC* (2011), p. 2; Eman Ghoneim, 'A Remote Sensing Study of Some Impacts of Global Warming on the Arab Region', in Tolba and Saab (eds), *Climate Change*, p. 36; GSDP, *Second HDR*, p. 112.

2. Richer, *Conservation*, pp. 6–7; Al Mulla, *Climate Change*, p. 14; MoEnv Qatar, *National Communication*, p. 1.

3. GSDP, *National Vision*, p. 30.

4. GSDP, *Second HDR*, pp. 111–12.

5. Al Mulla, *Climate Change*, pp. 15–16; quoted in *The Peninsula*, 28 January 2009.

6. MoEnv Qatar, *National Communication*, pp. 2, 10.

7. In 1997, Kassler and Paterson predicted GDP losses of close to 5% of business as usual for Qatar by 2010. Peter Kassler and Matthew Paterson, *Energy Exporters and Climate Change* (London: Royal Institute of International Affairs, 1997), p. 38; Barnett *et al.*, 'OPEC Lose?', pp. 2084–5; Chatham House, *OPEC*, p. 20. The OPEC's own model from the late 1990s suggested that Qatar would be the most vulnerable of all member states to the negative impacts of response measures, with GDP losses of 3.3% (US$400m) in 2010. Similarly, a Chatham House report from 2005 ranked Qatar among the most vulnerable oil exporting countries to the impacts of Kyoto Protocol implementation, owing to the country's high dependence on energy exports.

8. Ahmed Azhari and Mohammed Al Maslamani, 'Anticipated Economic Costs and Benefits of Ratification of the Kyoto Protocol by the State of Qatar', *Climate Policy*, 4 (2004), *passim*.

9. Al Mulla, *Climate Change*, pp. 17–18, quoting, *inter alia* Barnett and Dessai, 'Articles 4.8 and 4.9', p. 234.

10. Barnett *et al.*, 'OPEC Lose?', p. 2086; Chatham House, *OPEC*, p. 20; Ahmed and Maslamani, 'Anticipated'; Al Mulla, *Climate Change*, p. 18.

11. MoEnv Qatar, *National Communication*, p. 1.

12. GSDP, *Second HDR*, p. 112.

13. Ibid., pp. 40–2.

14. Lower estimate: Nimah, 'Water Resources', p. 65. Higher estimate: GSDP, *Second HDR*, p. 39.

15. Richer, *Conservation*, 9; GSDP, *Second HDR*, p. 49.

16. Mohammed A. Dawoud, *et al.*, 'Using Renewable Energy Sources for Water Production in Arid Regions: GCC Countries Case Study', in A.M.O. Mohamed (ed.), *Arid Land Hydrogeology: In Search of a Solution to a Threatened Resource*, Vol. IV, DARE Series (London: Taylor & Francis, 2006), p. 121; Richer, *Conservation*, p. 6; FAO, *Aquastat online*, accessed in November 2010; GSDP, *Second HDR*, p. 44.

17. GSDP, *Second HDR*, pp. 40–3; IMF, *Qatar: Statistical Appendix*; World Bank, *A*

Water Sector Assessment Report on the Countries of the Cooperation Council of the Arab States of the Gulf, Report No. 32539-MNA (2005), p. 13.

18. United Nations, *Qatar Country Profile*, p. 23; GSDP, *Second HDR*, p. 46; *Gulf News*, 12 July 2010.

19. *MEED*, 5–11 March 2010; *Zawya Dow Jones*, 3 March 2011.

20. World Bank, *Water Assessment*, p. 22.

21. *Qatar Tribune*, 29 June 2010.

22. *MEED*, 5–11 March 2010.

23. Fahad al-Attiyah, QNFSP, presentation in a CIRS working group meeting, Georgetown University in Qatar, 13 November 2011.

24. United Nations, *Qatar Country Profile*, p. 23.

25. Al Mulla, *Climate Change*, p. 16; GSDP, *Second HDR*, pp. 48–9; QNFSP website, http://www.qnfsp.gov.qa/, accessed in January 2012.

26. Al Mulla, *Climate Change*, p. 16; UNEP, *Land-Based*, p. 47.

27. EIU, *Qatar: Country Profile*, p. 17.

28. Al Mulla, *Climate Change*, p. 16; GSDP, *Second HDR*, p. 48; Richer, *Conservation*, pp. 6, 8.

29. Gulf Research Center, *Green Gulf*, p. 52; United Nations, *Qatar Country Profile*, p. 23.

30. Al Mulla, *Climate Change*, p. 16.

31. *MEED*, 5–11 March 2010; Fahad al-Attiyah, Georgetown University in Qatar, 13 November 2011.

32. Hassan Al Mohannadi *et al.*, 'Residential Water Demand in Qatar, An Assessment', *Ambio* 32 (2003), p. 364.

33. Ibid., p. 363; GSDP, *Second HDR*, p. 48.

34. FAO, *Aquastat online*, accessed in November 2010; GSDP, *Development Strategy*, p. 217; *MENAFN*, 19 March 2009.

35. GSDP, *Development Strategy*, p. 218. OECD data from World Bank.

36. Al Mohannadi *et al.*, 'Residential Water', p. 363.

37. GSDP, *Second HDR*, p. 48; GSDP, *Development Strategy*, p. 219.

38. GSDP, *Second HDR*, p. 61.

39. Al Mohannadi *et al.*, 'Residential Water', p. 364; GSDP, *Development Strategy*, pp. 81, 84.

40. GSDP, *Second HDR*, p. 55; *The Peninsula*, 23 November 2006.

41. GSDP, *Development Strategy*, pp. 82–3, 217.

42. GSDP, *Second HDR*, pp. 21, 61.

43. *UAE Interact*, 22 March 2009; GSDP, *Second HDR*, p. 43; GSDP, *Development Strategy*, p. 219.

44. GSDP, *Development Strategy*, pp. 82–4.

45. *MEED*, 5–11 March 2010.

46. GSDP, *Development Strategy*, p. 43.

47. *Arabian Business*, 2 June 2010.

48. Summary on QNFSP website: http://www.qnfsp.gov.qa/, accessed in January 2011.

49. EIU, *Resources for the Future*, p. 16.
50. GSDP, *Second HDR*, p. 55; United Nations, *Qatar Country Profile*, p. 24.
51. *Financial Times*, 19 November 2009.
52. Hassad Food website: http://www.hassad.com/, accessed in August 2010.
53. Braun and Meizen-Dick, *Land grabbing*, p. 2; *The National*, 5 June 2009. No follow-up, however, ensued, and a leaked US embassy cable has later suggested that China, a rising actor in Africa, had sidelined Qatar as the preferred partner in the 'black box' negotiations of Kenyan President Kibaki's office. US Embassy in Nairobi, 'Chinese Engagement in Kenya' (17 February 2010), via http://www.cablesearch.org/.
54. *Reuters*, 11 August 2009.
55. *AMEinfo*, 3 June 2009; *Australian Financial Review*, 8 March 2010; *QNA*, 22 December 2009.
56. *Gulf Times*, 15 October 2011.
57. QNFSP brochure, undated.
58. *Gulf Times*, 12 December 2011.
59. QNFSP presentation in SustainableQatar's monthly lecture, Doha, 6 December 2011. Parts published in a summary by the author on SustainableQatar's website.
60. QNFSP brochure, undated, p. 11.
61. *The Daily Star Lebanon*, 5 January 2012.
62. WRI, *CAIT 9.0*.
63. Energy sector: 92% of total CO_2 and N_2O and 81% of CH_4 emissions. MoEnv Qatar, *National Communication*, pp. 19–25.
64. GSDP, *Development Strategy*, p. 223.
65. E.g. GSDP, *Second HDR*, p. 105. Trains: US EIA, *Qatar: Country Analysis Brief* (2011).
66. Al Mulla, *Climate Change*, pp. 9–11; MoEnv Qatar, *National Communication*, p. 25.
67. *The News Flare*, No. 8 (2009), p. 3; GSDP, *Second HDR*, p. 118.
68. MoEnv Qatar, *National Communication*, p. 3.
69. Fenhann, 'CDM pipeline', updated 1 November 2011.
70. *Gulf Times*, 19 September 2010.
71. Al-Othman, *Bare Hands*, pp. 129, 132.
72. *The National*, 15 April 2010.
73. OPEC, Statement to the United Nations Climate Change Conference (COP13), Bali, 14 December 2007; Carbon Trust press release, 2 November 2008.
74. *Gulf Times*, 26 February 2010; 12 December 2011.
75. *Gulf Times*, 18 November 2009.
76. GSDP, *Development Strategy*, p. 222.
77. Qatar Cool website: http://www.qatarcool.com/, accessed in January 2011.
78. *The National*, 29 April 2010.
79. Qatar National Bank, *Economic Insight*, p. 18.
80. GSDP, *Development Strategy*, p. 224.

81. GSDP, *Second HDR*, p. 116.
82. MoEnv Qatar, *National Communication*, pp. 2–3.
83. Al Mulla, *Climate Change*, pp. 12–14; GSDP, *Second HDR*, pp. 121, 124.
84. Conversations in Doha, autumn 2011.
85. Qatar Foundation website: http://www.qf.org.qa/, accessed in August 2010.
86. Kamrava, 'Factionalism', pp. 407, 415.
87. Kéchichian, *Power*, p. 217; Kamrava, 'Factionalism', p. 407.
88. Qatar Foundation, website: http://www.qf.org.qa/, accessed in August 2010.
89. See e.g. Virginia Commonwealth University press release, 25 June 2002.
90. MEED website: http://www.meed.com/knowledge-bank/top-100-projects/, accessed in January 2012; Qatar Foundation, *Annual Report 2009–2010* (undated), p. 17.
91. QSTP press release, 22 February 2009.
92. Qatar Foundation website: http://www.qf.org.qa/, accessed in August 2010; QSTP, 'Press pack', http://www.qstp.org.qa/files/pdf/QSTPPressPack.pdf, accessed in August 2010.
93. Tidu Maini, speech at the *All Energy Conference*, Aberdeen, 20 May 2009.
94. Interview with Dr Eulian Roberts, QSTP, Doha, November 2010.
95. QSTP press release, 22 February 2009; QSTP, 'Press pack'; QNA, 22 February 2012.
96. Maini, speech, Aberdeen, 20 May 2009; Imperial College press release, 18 September 2001; QSTP website: http://www.qstp.org.qa/, accessed in August 2010.
97. Imperial College press release, 9 June 2008; quote from Imperial College Qatar Carbonates and Carbon Storage Research Centre website: http://www3.imperial.ac.uk/qatarcarbonatesandcarbonstorage, accessed in August 2010.
98. Brookings Institute, *Carbon Capture and Sequestration*, Doha Carbon and Energy Forum Briefing Paper (2010), p. 2.
99. Chevron Center for Sustainable Energy Efficiency website: http://www.chevroncsee.com/, accessed in January 2012.
100. GreenGulf press release, 18 March 2009.
101. Correspondence with Omran Al Kuwari, GreenGulf, February 2011; QSTP press Release, 25 April 2010; GreenGulf press release, 24 April 2010.
102. 1% owned by Qatar National Bank. Interview with Craig Field and Narasimha Raghavan, Qatar Solar Technologies, Doha, November 2010; Qatar Foundation website: http://www.qf.org.qa/, accessed in January 2012.
103. Maini, speech, Aberdeen, 20 May 2009.
104. QSTP press release, 7 October 2008.
105. Interview with Yousef Al Horr, Doha, November 2010; *Construction Week Online*, 10 April 2009; QSTP press release, 17 June 2009.
106. Interview with Yousef Al Horr, Doha, November 2010; *Gulf Times*, 18 November 2009; BQDRI, *Lusail: QSAS Fact Sheet* (June 2010).
107. BQDRI press release, 16 March 2010; 27 May 2010; 30 May 2010; *Qatar Construction Sites*, issue 55, p. 7.

108. GORD press release, 17 June 2012.
109. Global Water Sustainability Center website: http://www.globalwsc.com/, accessed in January 2012.
110. Maini, speech, Aberdeen, 20 May 2009; QSTP press release, 31 May 2004; 2 April 2008; 11 April 2010.
111. Texas A&M Qatar website: http://meen.qatar.tamu.edu/, accessed in August 2010.
112. *The Foundation*, Issue 16 (2010), p. 5.
113. Nidhi Kalra *et al.*, *Recommended Research Priorities for the Qatar Foundation's Environment and Energy Research Institute*, RAND-Qatar Policy Institute (2011), *passim*.
114. Conversations with stakeholders in Education City, autumn 2011 and January 2012.
115. *Gulf Times*, 26 February 2010.
116. *MEED*, 18 September 2009.
117. Conversation with stakeholder in Education City, Doha, January 2012; *The National*, 15 April 2010.
118. *Gulf Times*, 9 September 2009.
119. *Construction Week Online*, 25 May 2009.
120. Correspondence with Christopher Silva, Qatar Foundation, January 2012.
121. *Railway Gazette*, 30 July 2012.
122. Maini, speech, Aberdeen, 20 May 2009.
123. Qatar Science and Technology Park, 'Press pack'.
124. *Arabian Business*, 9 June 2009.
125. Term used by the QSTP. Interview with Eulian Roberts, Doha, November 2010.
126. Interview with Qatar-based renewable energy experts, Doha, November 2010.
127. Interviews with Qatar-based technology expert and renewable energy experts, Doha, November 2010.
128. *Gulf Times*, 26 April 2010.
129. Interview with Eulian Roberts, Doha, November 2010; *SciDevNet*, 14 March 2007.
130. Maini, speech, Aberdeen, 20 May 2009; Omran al-Kuwari quoted in *GreenGulf* (24 April 2010).
131. Tidu Maini quoted in *Gulf Times*, 15 October 2009.
132. See e.g. Imperial College Qatar Carbonates and Carbon Storage Research Centre website: http://www3.imperial.ac.uk/qatarcarbonatesandcarbonstorage, accessed in August 2010; BQDRI website: http://www.bqdri.org/, accessed in September 2010.
133. GreenGulf press release, 18 March 2009; Yousef Al Horr quoted in BQDRI press release, 18 November 2009.
134. Msheireb Properties press release, 13 January 2010.
135. H.H. Sheikha Moza bint Nasser's website: http://www.mozabintnasser.qa/, accessed July 2012.

136. Steffen Hertog, *Princes, Brokers, and Bureaucrats: Oil and the State in Saudi Arabia* (Ithaca, NY: Cornell University Press, 2010), e.g. p. 56.

137. Correspondence, August 2010.

138. E.g. interview with Qatar-based renewable energy expert, Doha, November 2010.

8. THE GCC STATES IN THE INTERNATIONAL CLIMATE REGIME

1. Parts of earlier versions of the following subchapters have been published in Luomi, 'Gulf of Interest'.

2. Years of accession/ratification of UNFCCC/KP: Bahrain: 1994/2006; Kuwait: 1994/2005; Oman: 1995/2005; Qatar: 1996/2005; Saudi Arabia: 1994/2005; UAE: 1995/2005.

3. Joanna Depledge, 'Striving for No: Saudi Arabia in the Climate Change Regime', *Global Environmental Politics*, 8 (2008), p. 17.

4. Interview, Bonn UNFCCC climate conference, Bonn, June 2009.

5. For reasons including Shia populations and a jointly owned oilfield, and as showed by the events of 2011.

6. Interview with UAE climate policy expert, Abu Dhabi, October 2010.

7. Interview with Ibrahim Ahmed Al Ajmi, Ministry of Environment and Climate Affairs of Oman, Poznan, December 2008.

8. E.g. interviews with UAE climate policy experts and Qatar climate policy expert, Abu Dhabi and Doha, October 2009.

9. Chatham House, *OPEC*, p. 10.

10. Depledge, 'Striving', pp. 15, 17–18; Chatham House, *OPEC*, pp. 6, 9; Suraje Dessai, *An Analysis of the Role of OPEC as a G77 Member at the UNFCCC*, Report for WWF (2004), p. 16.

11. Mari Luomi, *Bargaining in the Saudi Bazaar: Common Ground for a Post-2012 Climate Agreement?* FIIA Briefing Paper, No. 48 (2009), p. 6.

12. Correspondence with Arab climate policy expert, April 2010.

13. Interviews with Wael Hmaidan, IndyACT, Poznan, December 2008; with small Gulf state's climate negotiator, Abu Dhabi, May 2010; personal correspondence, July 2012.

14. Based on an analysis in Mari Luomi, *Ilmasto- vai öljypolitiikkaa? Lähi-idän arabimaiden ilmastopolitiikan selitysten jäljillä*, FIIA Working Paper, No. 62 (2009), pp. 18–19.

15. Based on the official lists of participants to the Conferences of Parties 9–16 of the UNFCCC.

16. Interviews with UAE climate policy and environmental experts, Abu Dhabi, October 2009.

17. Presentations by Katherine Watts and Wael Hmaidan, NGO side-event, UNFCCC COP14, Poznan, 11 December 2008; Jon Barnett, 'The Worst of Friends: OPEC and G77 in the Climate Regime', *Global Environmental Politics*, 8 (2008), pp. 5–6.

18. Interview with Kati Kulovesi, Earth Negotiations Bulletin, Poznan, December 2008.
19. Based on appearance count in the International Institute for Sustainable Development (IISD) *Earth Negotiations Bulletins* in 1996–2008. Nb. The bulletins only cover the most important aspects of the negotiations.
20. Kassler and Paterson, *Energy Exporters*, p. 87; Aarts and Janssen, 'Shades of Opinion', p. 337.
21. Based on archival records and personal observation and interviews.
22. Parts of this section have been previously published: in Luomi, *Bargaining*.
23. WRI, *CAIT 8.0.*
24. Depledge, 'Striving', pp. 9–11, 17. Saudi Arabia acceded to the UNFCCC in 1994.
25. E.g. Dessai, *Role of OPEC*, pp. 20, 26; interview with Kati Kulovesi, Poznan, December 2008.
26. Antto Vihma, *Arrested Development*, FIIA Comment, No. 8/2011 (2011), p. 1.
27. Depledge, 'Striving', pp. 11, 20; Barnett, 'Worst of Friends', pp. 2–4.
28. Kassler and Paterson: *Energy Exporters*, pp. 98–9; Chatham House, *OPEC*, p. 7.
29. E.g. Dessai, *Role of OPEC*, p. 3.
30. E.g. interview with Kati Kulovesi, Poznan, December 2008. See also Heather McGray, 'From Copenhagen to Cancun: Adaptation', World Resources Institute, http://www.wri.org/stories/2010/05/copenhagen-cancun-adaptation/, dated 13 May 2010.
31. Presidency of Meteorology and Environment of Saudi Arabia, *First National Communication: Kingdom of Saudi Arabia* (2005), p. 121.
32. Barnett *et al.*, 'OPEC Lose?', p. 2086.
33. Luomi, *Bargaining*.
34. Irja Vormedal, 'The Influence of Business and Industry NGOs in Negotiation of the Kyoto Mechanisms: the case of Carbon Capture and Storage in the CDM', *Global Environmental Politics*, 8 (2008), p. 52.
35. Economic mitigation potential: 200–2,000 Gt by 2100, according to the IPCC, as cited in OPEC, *World Oil Outlook* (2008), pp. 8–9, 45–7.
36. Climate science and future predictions on emission trajectories show that the developed countries alone cannot prevent dangerous climate change even by cutting their emissions to zero.
37. Joanna Lewis and Elliot Diringer, *Policy-Based Commitments in a Post-2012 Climate Framework* (Arlington: Pew Center on Global Climate Change, 2007), p. 1.
38. Interview with UAE climate policy expert, Abu Dhabi, October 2009.
39. UNFCCC lists of participants for COPs in 2002–2010.
40. Interviews with Qatar and UAE climate policy experts, Doha, October 2009; Abu Dhabi, October 2010.
41. Based on personal observation, archives of the IISD *Earth Negotiations Bulletins* on UNFCCC meetings in 1995–2008, and official lists of participants from Conferences of Parties in 2000–10.

42. US EIA, *Saudi Arabia Country Analysis Brief* (January 2011).

43. Mari Luomi, *Oil or Climate Politics? Avoiding a Destabilising Resource Split in the Arab Middle East*, FIIA Briefing Paper, No. 58 (2010), p. 8

44. E.g. Latin American countries and environmental NGOs. Vormedal, 'CCS in the CDM', pp. 52–3.

45. *Bloomberg*, 30 September 2011.

46. Parts of this subchapter have been published in Luomi, 'Abu Dhabi's Alternative-Energy Initiatives', pp. 112–15.

47. World Council for Renewable Energy website: http://www.wcre.de/, accessed in January 2011.

48. See e.g. *BusinessGreen*, 26 January 2009.

49. Preparatory Commission for IRENA, *Report of the First Session of the Preparatory Commission*, IRENA/PC.1/SR (27 January 2009), p. 3.

50. *Gulf News*, 29 June 2009; *The National*, 10 June 2009.

51. *The National*, 30 June 2009.

52. *Gulf News*, 24 June 2009; *The National*, 29 June 2009.

53. US Embassy in Accra, 'Demarche on Delivery on International Renewable Energy Agency' (17 June 2009), via http://www.cablesearch.org/.

54. Sheikha Lubna al-Qasimi, quoted in *Gulf News*, 19 June 2009.

55. IRENA@UAE, 'The Future is Here' presentation, http://www.irenauae.com/, accessed on 15 July 2009; *Gulf News*, 1 May 2009; *The National*, 19 April 2009; 6 May 2009; 21 June 2009; 1 July 2009.

56. WAM, 18 June 2009.

57. *The National*, 30 June 2009; 10 July 2009.

58. *MEED*, 1 July 2009.

59. Interview with UAE climate policy expert, Abu Dhabi, October 2010; *Gulf News*, 28 June 2009.

60. US Embassy in Abu Dhabi, 'FM Presses SRAP Holbrooke for IRENA Support' (25 March 2009), via http://www.cablesearch.org/.

61. *Gulf News*, 28 June 2009.

62. IRENA@UAE website: http://www.irenauae.com/, accessed in August 2009.

63. *The National*, 8 July 2009.

64. Interviews with UAE climate policy experts, Abu Dhabi, October 2010.

65. US Embassy in Abu Dhabi, 'UAE Asks to Be Major Economies Forum Observer' (2 April 2009), via http://www.cablesearch.org/.

66. Personal observation based on conversation in a Ministry of Foreign Affairs seminar, Abu Dhabi, May 2010.

67. US Embassy in Abu Dhabi, 'Progress Made with the International Renewable Energy Agency, but Long Road ahead' (21 February 2010), via http://www.cablesearch.org/.

68. According to a plan from 2010, 42% in 2011. *The National*, 20 October 2010.

69. Personal conversations, Abu Dhabi, 2009–2010.

70. Mohamed Raouf, presentation in seminar on *Arab World Policy for post-2012 Negotiations*, Beirut, 15 October 2008.

71. See e.g. EAD, *Policies and Regulations*, p. 20; EAD press release, 22 January 2008; Majid Al Mansouri, presentation at the First International Conference on the Clean Development Mechanism, Saudi Arabia, 19–21 September 2006.

72. Al Mansouri presentation, Saudi Arabia, 2006; CDM-DNA UAE website: http//www.cdm-uae.ae/, accessed in December 2010; UNFCCC website: http://unfccc.int/, accessed in January 2011.

73. EAD, *Policies and Regulations*, p. 20; interview with UAE climate change expert, Abu Dhabi, October 2010.

74. UAE Ministry of Foreign Affairs website: http://www.mofa.gov.ae/, accessed in January 2012.

75. Interviews with UAE climate policy experts, Abu Dhabi, October 2010.

76. UNFCCC website: http://unfccc.int/, accessed in December 2010.

77. Based on official lists of participants of COPs 1–16 (1995–2010).

78. Official lists of participants of the UNFCCC.

79. Contact group co-chair and COP bureau vice president, one workshop. IISD, *Earth Negotiations Bulletin*, 12 (2003, 2006); *AMEinfo*, 4 September 2006.

80. IISD, *Earth Negotiations Bulletin*, 12 (1996–2010).

81. Interview with Rashid Ahmed bin Fahad, Minister of Environment and Water of the UAE, Poznan, December 2008.

82. Interviews with UAE climate policy and environmental experts, Abu Dhabi and Dubai, October 2009.

83. UNFCCC website: http://unfccc.int/, accessed in January 2011.

84. Speech in joint high-level segment of COP and CMP, 15th Conference of the Parties, UNFCCC, December 2010.

85. E.g. speech by Sheikh Abdullah bin Zayed Al Nahyan, New York, 24 September 2007; Mohammed bin Dhaen al-Hamli in *UAE Interact*, 3 February 2009; 23 April 2009; Rashid bin Fahad in: *The National*, 23 April 2009.

86. Personal observation based on conversation in a Ministry of Foreign Affairs seminar, Abu Dhabi, May 2010.

87. Interview with small Gulf states' climate negotiator, Abu Dhabi, May 2010.

88. Minister of State for Foreign Affairs of the UAE, Letter to Yvo de Boer, Ref. 3784 (14 February 2010).

89. Based on interview with UAE climate policy expert, Abu Dhabi, October 2010.

90. Statement in joint high-level segment of COP and CMP, 16th Conference of the Parties, UNFCCC, 9 December 2010.

91. *Arabian Business*, 17 January 2012.

92. *The National*, 19 June 2009.

93. United Nations press release ENV/DEV/1149, 9 August 2010.

94. *The National*, 26 June 2010; Clean Energy Ministerial website: http://www.cleanenergyministerial.org/, accessed in January 2011.

95. WREC website: http://www.wrenuk.co.uk/, accessed in January 2011; WFES press release, 30 October 2010.

96. Joint statement from the foreign ministers of the UAE, Cape Verde, Costa Rica, Iceland, Singapore and Slovenia, 9 December 2009.

97. World Future Energy Summit website: http://www.worldfutureenergysummit. com/, accessed in January 2011.

98. *The National*, 4 May 2009.

99. Interviews with Masdar expert and UAE climate policy expert, Abu Dhabi, October 2010; *The National*, 30 June 2009; IISD, *Earth Negotiations Bulletin*, 12:553 (2012).

100. Under-secretary general Khalid Ghanim Al Ali. UNFCCC website: http:// unfccc.int/, accessed in January 2011.

101. Al Shaheen was Qatar's only CDM project until April 2010 when a small waste heat project (7 kt/year) at Ras Laffan was submitted for validation. See e.g. *Gulf Times*, 2 June 2006; Fenhann, 'CDM pipeline', updated on 1 January 2011.

102. Presentation by Adnan Fahad Al-Ramzani, *Lessons Learned from Al-Shaheen (ALS) Oil Field, Gas Recovery and Utilization Project* in Environmental Challenges in Gas Processing, Doha, 5 November 2008.

103. *Gulf Times*, 29 October 2007.

104. UNFCCC, *FCCC/SBSTA/2002/MISC.3*.

105. Al Mulla, *Climate Change*, pp. 6–7; Abdullah al-Midhadi's interview in: *RasGas Magazine*, issue 29 (2010), p. 9.

106. UNFCC website: http://unfccc.int/, accessed in December 2011.

107. Based on personal experience as a UNFCCC observer and fieldwork in Doha in 2008–11.

108. Interview with Qatari negotiator, Abu Dhabi, October 2009.

109. Based on official lists of participants to COPs 1996–2011.

110. In terms of submissions of views, Qatar has been the most active of the small GCC states, with seven documents submitted in 2001–2009. UNFCCC website: http://unfccc.int/, accessed in October 2010.

111. UNFCCC, *FCCC/CP/2001/MISC.1*, p. 104; UNFCCC, *FCCC/CP/2001/ MISC.1/Add.1*, p. 2.

112. UNFCCC, *FCCC/SBSTA/2002/MISC.3*, pp. 16–35.

113. UNFCCC, *FCCC/KP/CMP/2006/MISC.2*, pp. 31–2; UNFCCC, *FCCC/KP/ AWG/2009/MISC.4*, p. 12; UNFCCC, *FCCC/AWGLCA/2009/MISC.1/Add.1*, pp. 16–17; UNFCCC, *FCCC/AWGLCA/2009/MISC.4 (Part II)*, pp. 75–6; UNFCCC, *FCCC/SB/2011/MISC.1*, pp. 15–17, UNFCCC, *FCCC/AWGLCA/ 2011/CRP.29*; UNFCCC, *FCCC/SBSTA/2011/MISC.10*, pp. 53–4; IndyACT, *Arab Position Matrix*, 8 October 2009. Unpublished.

114. IISD, *Earth Negotiations Bulletin*, 12 (1997–2011).

115. IISD, *Earth Negotiations Bulletin* (2008–2010).

116. IISD, *Earth Negotiations Bulletin* (1999; 2010).

117. Al Mulla, *Climate Change*, pp. 7, 18–19.

118. Speeches in English are only available for COP8–9; COP15; COP17.

119. *Bloomberg*, 14 October 2009; *Gulf Times*, 21 April 2008.

120. Interview with Qatar-based renewable energy expert, Doha, 4 November 2010.

121. *Bloomberg*, 14 October 2009; *The Peninsula*, 11 September 2009; *The National*, 19 January 2010.

122. Sheikh Hamad bin Khalifa Al Thani's statements in the General Debate of the 62nd Session of the UNGA, New York, 25 September 2007; the General Debate of the 64th Session of the UNGA, New York, 23 September 2009.

123. Abdullah Al Midhadi in Oxford Business Group: *The Report: Qatar 2009*, p. 239; GSDP, *National Vision*, p 33; GSDP, *Second HDR*, p. 123.

124. Based on IISD's daily negotiating summaries.

125. UNFCCC, website: http://unfccc.int/, accessed in October 2010.

126. *Guardian*, 29 November 2011; *New York Times*, 31 May 2011.

127. Correspondence with Qatar-based foreign policy expert, January 2012.

128. Conversations in Abu Dhabi, December 2011.

129. Statement in joint high-level segment of COP and CMP, 17th Conference of the Parties, UNFCCC, 7 December 2011.

130. *The Peninsula*, 12 May 2012.

SELECTED BIBLIOGRAPHY

Books, book chapters and scientific journal articles

Aarts, Paul and Dennis Janssen, 'Shades of Opinion: The Oil Exporting Countries and International Climate Politics', *The Review of International Affairs*, 3 (2003), pp. 332–51.

Al-Iriani, Mahmoud A., 'Climate-Related Electricity Demand-Side Management in Oil-Exporting Countries—The Case of the United Arab Emirates', *Energy Policy*, 33 (2004), pp. 2350–60.

Al Mohannadi, Hassan, Chris Hunt and Adrian Wood, 'Residential Water Demand in Qatar, An Assessment', *Ambio*, 32 (2003), pp. 326–55.

Al-Othman, Nasser, *With their Bare Hands: The Story of the Oil Industry in Qatar* (Harlow: Longman, 1984).

Al Shamsi, Fatima, 'Industrial Strategies and Change in the UAE during the 1980s', in Abbas Abdelkarim (ed.), *Change and Development in the Gulf* (Basingstoke and London: Macmillan, 1999), pp. 79–103.

Anderson, Lisa, 'Dynasts and Nationalists: Why Monarchies Survive', in Joseph Kostiner (ed.), *Middle East Monarchies The Challenge of Modernity* (Boulder and London: Lynne Rienner, 2000), pp. 53–69.

Aspinall, Simon, 'Environmental Development and Protection in the UAE', in I. Al Abed and P. Hellyer (eds), *United Arab Emirates: A New Perspective* (London: Trident Press, 2001), pp. 277–304.

Azhari, Ahmed and Mohammed Al Maslamani, 'Anticipated Economic Costs and Benefits of Ratification of the Kyoto Protocol by the State of Qatar', *Climate Policy*, 4 (2004), pp. 75–80.

Barnett, Jon, 'The Worst of Friends: OPEC and G77 in the Climate Regime', *Global Environmental Politics*, 8 (2008), pp. 1–8.

Barnett, Jon and Suraje Dessai, 'Articles 4.8 and 4.9 of the UNFCCC: Adverse Effects and the Impacts of Response Measures', *Climate Policy*, 2 (2002), pp. 231–9.

Barnett, Jon, Dessai Suraje and Michael Webber, 'Will OPEC Lose from the Kyoto Protocol?', *Energy Policy*, 32 (2004), pp. 2077–88.

Brown, Gavin (ed.), *OPEC and the World Energy Market: A Comprehensive Reference Guide* (Harlow: Longman, 1991).

Butt, Gerald, 'Oil and Gas in the UAE', in Ibrahim Al Abed and Peter Hellyer (eds), *United Arab Emirates: A New Perspective* (London: Trident Press, 2001), pp. 231–48.

Coates Ulrichsen, Kristian, 'Internal and External Security in the Arab Gulf States', *Middle East Policy*, 16 (2009), pp. 39–58.

Cordesman, Anthony H., *Energy Developments in the Middle East* (Westport: Praeger, 2004).

Crystal, Jill, *Oil and Politics in the Gulf: Rulers and Merchants in Kuwait and Qatar* (Cambridge University Press, 1995).

Davidson, Christopher, *The United Arab Emirates: A Study in Survival* (Boulder: Lynne Rienner, 2005).

———, 'The Emirates of Abu Dhabi and Dubai: Contrasting Roles in the International System', *Asian Affairs*, 38 (2007), pp. 33–48.

———, *Dubai: The Vulnerability of Success* (London: Hurst, 2008).

———, *Abu Dhabi: Oil and Beyond* (London: Hurst, 2009).

———, 'Abu Dhabi's New Economy: Oil, Investment and Domestic Development', *Middle East Policy*, 16 (2009), pp. 59–79.

Dawoud, Mohammed A., A.R. Allam, M.A. El Shewey and S.M. Soliman, 'Using Renewable Energy Sources for Water Production in Arid Regions: GCC Countries Case Study', in A.M.O. Mohamed (ed.), *Arid Land Hydrogeology: In Search of a Solution to a Threatened Resource*, Vol. IV, DARE Series (London: Taylor & Francis, 2006), pp. 117–29.

Depledge, Joanna, 'Striving for No: Saudi Arabia in the Climate Change Regime', *Global Environmental Politics*, 8 (2008), pp. 9–35.

Doukas, Haris, Konstantinos D. Palitzianas, Argyris G. Kagiannias and John Psarras, 'Renewable Energy Sources and Rationale Use of Energy Development in the Countries of GCC: Myth or Reality?', *Renewable Energy* 31 (2006), pp. 755–70.

Emirates Center for Strategic Studies and Research, *With United Strength: H.H. Shaikh Zayid Bin Sultan Al Nahyan, The Leader and the Nation* (Abu Dhabi: ECSSR, 2004).

Gause, Gregory, 'The Persistence of Monarchy in the Arabian Peninsula: A Comparative Analysis', in Joseph Kostiner (ed.), *Middle East Monarchies: The Challenge of Modernity* (Boulder and London, Lynne Rienner, 2000), pp. 167–86.

Ghanem, Shokri, Rezki Lounnas and Garry Brennand, 'The Impact of Emissions Trading on OPEC', *OPEC Review*, 23 (1999), pp. 79–112.

Giddens, Anthony, *The Politics of Climate Change* (Cambridge: Polity Press, 2009).

Harry, Wes, 'Employment Creation and Localisation: The Crucial Human Resource Issues for the GCC', *International Journal of Human Resource Management*, 18 (2007), pp. 357–75.

Heard-Bey, Frauke, 'The United Arab Emirates: Statehood and Nation-Building in a Traditional Society', *Middle East Journal*, 59 (2005), pp. 357–75.

Herb, Michael, *All in the Family: Absolutism, Revolution, and Democracy in the Middle Eastern Monarchies* (Albany: State University of New York, 1999).

Hertog, Steffen, *Princes, Brokers, and Bureaucrats: Oil and the State in Saudi Arabia* (Ithaca, NY: Cornell University Press, 2010).

Hinnebusch, Raymond, *The International Politics of the Middle East* (Manchester University Press, 2003).

Hudson, Michael C., *Arab Politics: The Search for Legitimacy* (New Haven and London: Yale University Press, 1977).

Hvidt, Martin, 'The Dubai Model: An Outline of Key Development-Process Elements in Dubai', *International Journal of Middle East Studies*, 41 (2009), pp. 397–418.

Kadhim, Abbas, 'The Future of Nuclear Weapons in the Middle East', *Nonproliferation Review*, 13 (2006), pp. 581–9.

Kamrava, Mehran, 'Royal Factionalism and Political Liberalization in Qatar', *Middle East Journal*, 63 (2009), pp. 401–20.

Kassler, Peter and Matthew Paterson, *Energy Exporters and Climate Change* (London: Royal Institute of International Affairs, 1997).

Karl, Terry Lynn, 'The Perils of the Petro-State: Reflections on the Paradox of Plenty', *Journal of International Affairs*, 53 (1999), pp. 31–48.

Kazim, Ayoub M., 'Assessments of Primary Energy Consumption and its Environmental Consequences in the United Arab Emirates', *Renewable and Sustainable Energy Reviews*, 11 (2007), pp. 426–46.

Kéchichian, Joseph A., *Power and Succession in Arab Monarchies: A Reference Guide* (London: Lynne Rienner, 2008).

Luomi, Mari, 'Abu Dhabi's Alternative-Energy Initiatives: Seizing Climate-Change Opportunities', *Middle East Policy*, 16 (2009), pp. 102–17.

———, Gulf of Interest: Why Oil Still Dominates Middle Eastern Climate Politics', *Journal of Arabian Studies*, 1 (2011), pp. 249–66.

———, 'The Economic and Prestige Aspects of Abu Dhabi's Nuclear Programme', in Mehran Kamrava (ed.), *The Nuclear Question in the Middle East* (London: Hurst, 2012), pp. 125–58.

Marcel, Valérie, *Oil Titans: National Oil Companies in the Middle East* (Baltimore: Brookings, 2006).

Nader, Sam, 'Paths to a Low-Carbon Economy—the Masdar Example', *Energy Procedia*, 1 (2009), pp. 3951–58.

Nakhleh, Emile A., '"The Creation of Qatar" by Rosemarie Said Zahlan', review, *International Journal of Middle East Studies*, 16 (1984), pp. 295–6.

Nonneman, Gerd (ed.), *Analyzing Middle East Foreign Policies and the Relationship with Europe* (Abingdon, Oxon: Routledge, 2005).

O'Brien, James, Ramin Keivani and John Glasson, 'Towards a New Paradigm in Environmental Policy Development in High-Income Developing

Countries: The Case of Abu Dhabi, United Arab Emirates', *Progress in Planning*, 68 (2007), pp. 201–56.

Ouis, Pernilla, 'Greening in the Emirates: The Modern Construction of Nature in the United Arab Emirates', *Cultural Geographies*, 9 (2009), pp. 334–47.

Paterson, Matthew, *Global Warming and Global Politics* (London: Routledge, 1996).

Persson, Tobias A., C. Azar, D. Johansson and K. Lindgren, 'Major Oil Exporters May Profit Rather than Lose in a Carbon-Constrained World', *Energy Policy*, 32 (2007), pp. 6346–53.

Peterson, J.E., 'Qatar and the World: Branding for a Micro-State', *Middle East Journal*, 60 (2006), pp. 732–48.

Rabi, Uzi, 'Qatar's Relations with Israel: Challenging Arab and Gulf Norms', *Middle East Journal*, 63 (2009), pp. 443–59.

Reich, Bernard (ed.), *Political Leaders of the Contemporary Middle East and North Africa: A Bibliographical Dictionary* (Westport, Connecticut: Greenwood Press, 1990).

Taelman, Elisa, 'Saadiyat Island Tourist Development Project: Dredging in an Ecologically Sensitive Area', *Terra et Aqua*, issue 116 (2009), pp. 3–11.

Tuchman Mathews, Jessica, 'Redefining Security', *Foreign Affairs*, 68 (1989), pp. 162–77.

Working papers, reports, analyses, briefs, and other journals

Aarts, Paul, *The Arab Oil Weapon: A One-Shot Edition*, Emirates Occasional Paper, No. 34 (ECSSR, 1999).

ADNOC, *Health, Safety & Environment Report 2008* (2009).

———, *Developing our Natural Resources Responsibility: Abu Dhabi National Oil Company (ADNOC) 2010 Sustainability Report* (2011).

Al Awad, Mouawiya and Carole Chartouni, *Explaining the Decline in Fertility among Citizens of the G.C.C. Countries: the Case of the U.A.E.*, ISER Working Paper, No. 1 (2010).

Bachellerie, Imen Jerid, *Renewable Energy Transition in the GCC: Finding the Right Paradigms*, GRC Analysis (2010).

Beeah, *Overview of the State of Environment in the Emirate of Sharjah, U.A.E.* (undated).

Blanchard, Christopher M. and Paul K. Kerr, *The United Arab Emirates Nuclear Program and Proposed U.S. Nuclear Cooperation*, Congressional Research Service (17 July 2009).

Booz & Company, 'Gas Shortage in the GCC: How to Bridge the Gap', Perspective (2010).

Braun, Joachim and Ruth Meinzen-Dick, *'Land grabbing' by Foreign Investors in Developing Countries: Risks and Opportunities*, IFPRI Policy Brief, No. 13 (2009).

Brookings Institute, *Carbon Capture and Sequestration*, Doha Carbon and Energy Forum Briefing Paper (2010).

Brown, Oli and Alec Crawford, *Rising Temperatures, Rising Tensions: Climate Change and the Risk of Violent Conflict in the Middle East*, International Institute for Sustainable Development (2009).

Buhaug, Halvald, Nils-Petter Gleditsch and Ole Magnus Theisen, *Implications of Climate Change for Armed Conflict*, Social Dimensions of Climate Change (Washington: World Bank, 2008).

Burke, Sharon, *Natural Security*, CNAS Working Paper (2009).

Bushnak, Adil A., 'Desalination' in Mohamed el-Ashry, Najib Saab and Bashar Zeitoon (eds), *Arab Environment: Water. Sustainable Management of a Scarce Resource*, Arab Forum for Environment and Development (2010).

Caton, Steven and Nader Ardalan, *New Arab Urbanism: The Challenge to Sustainability and Culture in the Gulf*, John F. Kennedy School of Government, Harvard University (2010).

Chaaban, Farid, 'Air Quality', in Mostafa K. Tolba and Najib W. Saab (eds), *Arab Environment: Future Challenges*, Arab Forum for Environment and Development (2008), pp. 45–62.

Chalcraft, John, *Monarchy, Migration and Hegemony in the Arabian Peninsula*, LSE Kuwait Programme Research Paper No. 12 (2010).

Chatham House (Royal Institute of International Affairs, London), *OPEC and Climate Change: Challenges and Opportunities*, Briefing Paper (2005).

Chatila, Jean G., 'Municipal and Industrial Water Management', in Hussein Abaza *et al.* (eds), *Arab Environment: Green Economy. Sustainable Transition in a Changing Arab World*, Arab Forum for Environment and Development (2011), pp. 125–36.

CNA, *National Security and the Threat of Climate Change* (2007).

Cruz, R.V., H. Harasawa, M. Lal and S. Wu, 'Chapter 10: Asia', in M.L. Parry, O.F. Canziani, J.P. Palutikof, P.J. van der Linden and C.E. Hanson (eds), *Climate Change 2007: Impacts, Adaptation and Vulnerability*, Contribution of Working Group II to the Fourth Assessment Report of the Intergovernmental Panel on Climate Change (Cambridge University Press, 2007), pp. 469–506.

Darbouche, Hakim and Bassam Fattouh, *The Implications of the Arab Uprisings for Oil and Gas Markets*, Oxford Institute for Energy Studies, MEP 2 (2011).

Dargin, Justin, 'Qatar's Natural Gas: The Foreign-Policy Driver', *Middle East Policy*, 3 (2007), pp. 138–42.

———, *The Dolphin Project: The Development of a Gulf Gas Initiative*, Oxford Institute for Energy Studies, NG 22 (2008).

———, *Addressing the Natural Gas Crisis: Strategies for a Rational Energy Policy*, Belfer Center Dubai Initiative Policy Brief (2010).

Dasgupta, Susmita, Benoit Laplante, Craig Meisner, David Wheeler and Yan Jianping, *The Impact of Sea Level Rise on Developing Countries: A Comparative Analysis*, World Bank Policy Research Working Paper 4136 (2007).

Dawoud, Mohammed A., *Water Scarcity in GCC Countries: Challenges and Opportunities*, Gulf Research Center Research Paper (2007).

Dessai, Suraje, *An Analysis of the Role of OPEC as a G77 Member at the UNFCCC*, Report for WWF (2004).

Economist Intelligence Unit (EIU), *Oman: Country Profile 2008* (2008).

———, *United Arab Emirates: Country Profile 2008* (2008).

———, *Qatar: Country Profile 2009* (2009).

———, *United Arab Emirates: Country Report* (April 2009).

———, *United Arab Emirates: Country Report* (June 2009).

———, *Qatar: Energy Report*, EIU Industry Briefing (1 December 2009).

———, *Democracy Index 2010: Democracy in Retreat* (2010).

———, *The GCC in 2020: Resources for the Future* (2010).

———, *Qatar: Country Report* (January 2010).

———, *Qatar: Country Report* (December 2010).

———, *Qatar: Energy Report* (25 October 2011).

Environment Agency—Abu Dhabi (EAD), *Climate Change: Executive Summary* (2009).

———, *Climate Change: Impacts, Vulnerability & Adaptation* (2009).

Ecoventures, *The State of Environmental Initiatives Among UAE Companies* (2009).

El-Katiri, Laura, *Interlinking the Arab Gulf: Opportunities and Challenges of GCC Electricity Market Cooperation*, Oxford Institute of Energy Studies, EL 8 (2011).

EWS/WWF, *Dar al Khair*, 24 (2010).

EWS/WWF, Ministry of Environment and Water of the UAE, EAD and Global Footprint Network, *UAE Ecological Footprint Initiative* (2010).

Fattouh, Bassam, *The Drivers of Oil Prices: The Usefulness and Limitations of Non-Structural Model, the Demand–Supply Framework and Informal Approaches*, Oxford Institute for Energy Studies, WPM 32 (2007).

Fédération Internationale de Football Association, *2022 FIFA World Cup Bid Evaluation Report: Qatar* (2010).

Ghoneim, Eman, 'A Remote Sensing Study of Some Impacts of Global Warming on the Arab Region', in Mostafa K. Tolba and Najib W. Saab (eds), *Climate Change: Impact of Climate Change on Arab Countries*, Arab Forum for Environment and Development (2009), pp. 31–46.

Gulf Research Center, *Green Gulf Report* (2006).

———, *Gulf Yearbook 2006–2007* (2007).

Gulf States Newsletter, 'MBR means business in cabinet reshuffle', Issue 824 (February 2008).

Hertog, Steffen and Giacomo Luciani, *Energy and Sustainability Policies in the GCC*, LSE Kuwait Programme Working Paper, No. 6 (2009).

Hulbert, Matthew and Tariq Akbar, *Why a Gas Troika and Cartel Will Prove to be Hot Air*, Datamonitor Global Analysis (19 November 2008).

Intergovernmental Panel on Climate Change (IPCC), 'Summary for Policymakers', in M.L. Parry, O.F. Canziani, J.P. Palutikof, P.J. van der Linden and

C.E. Hanson (eds), *Climate Change 2007: Impacts, Adaptation and Vulnerability*, Contribution of Working Group II to the Fourth Assessment Report of the Intergovernmental Panel on Climate Change (Cambridge University Press, 2007), pp. 7–22.

International Energy Agency (IEA), *World Energy Outlook 2005: Middle East and North Africa Insights* (2005).

———, *Betwixt Petro-Dollars and Subsidies: Surging Energy Consumption in the Middle East and North Africa States*, IEA Information Paper (2008).

———, *World Energy Outlook 2008* (2008).

International Institute for Strategic Studies (IISS), *Nuclear Programmes in the Middle East: In the Shadow of Iran*, IISS Strategic Dossier (2008).

International Institute for Sustainable Development (IISD), *Earth Negotiations Bulletin*, 12 (1996–2011).

Kapiszewski, Andrzej, *Arab versus Asian Migrant Workers in the GCC Countries*, UN/POP/EGM/2006/02, UN Population Division (22 May 2006).

Khamis, May, Abdelhak Senhadji and team, *Impact of the Global Financial Crisis on the Gulf Cooperation Council Countries and Challenges Ahead* (Washington: IMF, 2010).

Launay, Frederic, *Environmental Situational Assessment for the GCC Countries*, GRC Research Paper (2006).

Luciani, Giacomo, 'The Gulf Countries and Nuclear Energy', *Gulf Monitor*, 6 (2007).

Luomi, Mari, *Bargaining in the Saudi Bazaar: Common Ground for a Post-2012 Climate Agreement?* FIIA Briefing Paper, No. 48 (2009).

———, *Ilmasto- vai öljypolitiikkaa? Lähi-idän arabimaiden ilmastopolitiikan selitysten jäljillä*, FIIA Working Paper, No. 62 (2009).

———, *Oil or Climate Politics? Avoiding a Destabilising Resource Split in the Arab Middle East*, FIIA Briefing Paper, No. 58 (2010).

Mitchell, John V. and Paul Stevens, *Ending Dependence: Hard Choices for Oil-Exporting States*, Chatham House Report (2008).

National Media Council, *UAE Yearbook 2009* (2009).

———, *UAE Yearbook 2010* (2010).

Nonneman, Gerd, *Political Reform in the Gulf Monarchies*, CMEIS Working Paper, University of Durham (2006).

OSEC, *Cleantech Business in the GCC: Market Assessment Report 2009* (2009).

Oxford Business Group, *The Report: Qatar 2010* (2010).

Partrick, Neil, *Nationalism in the Gulf States*, LSE Kuwait Programme Research Paper, No. 5 (2009).

Qatar Central Bank, *The Thirty Fourth Annual Report* (2011).

Qatar National Bank, *Qatar—Economic Insight* (September 2011).

Raouf, Mohamed A., *Economic Instruments as an Environmental Policy Tool: The Case of the GCC Countries*, GRC Gulf Paper (2007).

———, *Climate Change Threats, Opportunities, and the GCC Countries*, Middle East Institute Policy Brief, No. 12 (2008).

———, *Water Issues in the Gulf: Time for Action*, Middle East Institute Policy Brief, No. 22 (2009).

Richer, Renee, *Conservation in Qatar: Impacts of Increasing Industrialization*, CIRS Occasional Paper (2008).

Nimah, Musa N., 'Water Resources', in Mostafa K. Tolba and Najib W. Saab (eds): *Arab Environment: Future Challenges*, Report of the Arab Forum for Environment and Development (2008), pp. 63–74.

Qatar Foundation, *Annual Report 2009–2010* (undated).

Regulation and Supervision Bureau, *Electricity Tariffs for Large Users in the Emirate of Abu Dhabi*, Information Tariffs (November 2009).

———, *Water and Electricity Sector Overview 2008/2009* (2009).

———, *Annual Report 2010. For the Water, Wastewater and Electricity Sector in the Emirate of Abu Dhabi* (2011).

Rezai, Hamid, Simon Wilson, Michel Claereboudt and Bernard Riegl, 'Coral Reef Status in the ROPME Sea Area: Arabian/Persian Gulf, Gulf of Oman and Arabian Sea', in Clive Wilkinson (ed.), *Status of Coral Reefs of the World: 2004*, Vol. 1 (Australian Institute of Marine Sciences, 2004), pp. 155–70.

Setser, Brad and Rachel Ziemba, *GCC Sovereign Funds: Reversal of Fortune*, CFR Working Paper (2009).

Stern, Nicholas, *The Stern Review: The Economics of Climate Change* (Cambridge University Press, 2007).

The Foundation, Issue 16 (2010).

Tolba, Mostafa K. and Najib W. Saab (eds), *Arab Environment: Future Challenges*, Arab Forum for Environment and Development (2008).

UK Met Office, *Climate Change and the Middle East* (2009).

UK Trade & Investment, *Power & Water: Dubai and the Northern Emirates, United Arab Emirates (UAE)*, Sector Report (2009).

UN Division for Sustainable Development, *CSD-14/15 Thematic Profiles: Industry* (2007).

———, *CSD-14/15 Thematic Profiles: Atmosphere* (2007).

UNCTAD, *World Investment Report 2010. Investing in a Low-Carbon Economy* (2010).

UNEP, *Overview of Land-Based Sources and Activities Affecting the Marine Environment in the ROPME Sea Area*, UNEP Regional Seas Reports and Studies No. 168 (1999).

UNFPA, *State of the World Population 2011* (2011).

United Nations, *Johannesburg Summit 2002. Qatar Country Profile* (2002).

———, *The United Nations Regional Commissions and the Climate Change Challenges* (2009).

US Energy Information Administration (US EIA), 'Natural Gas Weekly Update' (1 December 2011).

Vihma, Antto, *Arrested Development*, FIIA Comment, No. 8/2011 (2011).

Wilén, Juha, *Arabiemiraatit Abu Dhabin rakentaminen* (Helsinki: Finpro toimialakatsaus, 2007).

Woertz, Eckart, 'Food Inflation in the GCC Countries', *Gulf Monitor*, 2 (2008).

———, 'The Gulf Food Import Dependence and Trade Restrictions of Agro Exporters in 2008', in S. Evenett (ed.), *Will Stabilisation Limit Protectionism? The 4th GTA Report* (London: Centre for Economic Policy Research, 2010), pp. 43–56.

World Bank, *A Water Sector Assessment Report on the Countries of the Cooperation Council of the Arab States of the Gulf*, Report No. 32539-MNA (2005).

World Commission on Environment and Development, *Our Common Future*, document A/42/427 (United Nations, 1987).

World Wildlife Fund for Nature (WWF), *No Energy Security without Climate Security* (2006).

———, *Living Planet Report 2010* (2010).

———, *Living Planet Report 2012* (2012).

Wright, Steven, *Generational Change and Elite-Driven Reforms in the Kingdom of Bahrain*, Durham Middle East Papers, No. 7 (2006).

Official documents

Abu Dhabi Urban Planning Council, *Plan Abu Dhabi 2030: Urban Structure Framework Plan* (2007).

Environment Agency—Abu Dhabi (EAD), *Policies and Regulations of Abu Dhabi Emirate, United Arab Emirates* (2008).

———, *Abu Dhabi Water Resources Master Plan* (2009).

———, *Entity Strategic Plan 2009–2013* (2009).

———, *Annual Report 2009–2010* (2010).

European Commission, *Green Paper: A European Strategy for Sustainable, Competitive and Secure Energy*, COM (2006) 105 final (8 March 2006).

General Secretariat for Development Planning (GSDP), *Qatar National Vision 2030* (2008).

———, *Second National Human Development Report. Advancing Sustainable Development: Qatar National Vision 2030* (2009).

———, *Qatar National Development Strategy 2011–16* (2011).

Government of Abu Dhabi, *The Abu Dhabi Economic Vision 2030*, Context and Executive Summary (2008).

Government of the UAE, *Policy of the United Arab Emirates on the Evaluation and Potential Development of Peaceful Nuclear Energy* (20 April 2008).

Intergovernmental Panel on Climate Change (IPCC), *Climate Change 2007: Synthesis Report. Summary for Policymakers* (November 2007).

Kingdom of Bahrain, *Bahrain's Initial National Communication to the UNFCCC. Volume I: Main Summary Report*, GCPMREW (2005).

Minister of State for Foreign Affairs of the UAE, Letter to Yvo de Boer, Ref. 3784 (14 February 2010).

Ministry of Economy of the UAE and UNDP, *Millennium Development Goals. United Arab Emirates*. Second Report (2007).

Ministry of Energy of the UAE (MoE-UAE), *The United Arab Emirates: Initial National Communication to the UNFCCC* (2006).

———, *Statistical Report 2003–2007* (2008), in Arabic.

———, *The United Arab Emirates: Second National Communications to the Conference of the Parties of UNFCCC* (2010).

Ministry of Environment (MoEnv) of Qatar, *Initial National Communication to the UNFCCC* (2011).

Preparatory Commission for IRENA, *Report of the First Session of the Preparatory Commission*, IRENA/PC.1/SR (27 January 2009).

Presidency of Meteorology and Environment of Saudi Arabia, *First National Communication: Kingdom of Saudi Arabia* (2005).

UN General Assembly, *Climate Change and Its Possible Security Implications*, A/63/L.8/Rev.1 (18 May 2009).

———, *Declaration on the Right to Development*, A/RES/41/128 (4 December 1986).

UN Security Council, 'Security Council holds first-ever debate on impact of climate change on peace, security, hearing over 50 speakers', SC/9000 (17 April 2007).

UNFCCC, FCCC/CP/2001/MISC.1.

———, FCCC/CP/2001/MISC.1/Add.1.

———, FCCC/SBSTA/2002/MISC.3.

———, FCCC/KP/CMP/2006/MISC.2.

———, FCCC/AWGLCA/2009/MISC.1/Add.1.

———, FCCC/AWGLCA/2009/MISC.4 (Part II).

———, FCCC/KP/AWG/2009/MISC.4.

———, FCCC/AWGLCA/2011/CRP.29.

———, FCCC/SB/2011/MISC.1.

———, FCCC/SBSTA/2011/MISC.10.

Unpublished documents

Al Mulla, Ali Hamed, *Chapter 4: Climate Change and Human Development in Qatar: Issues, Challenges and Opportunities*, Unpublished, drafted for Qatar's Second Human Development Report (2009).

———, 'Post 2012 Kyoto Protocol Climate Change Negotiations: Issues and Strategic Challenges to Qatar', paper presented at the *7th Natural Gas Conference*, Doha, 10–12 March 2009.

IndyACT, *Arab Position Matrix*, 8 October 2009.

Kumetat, Dennis, 'Climate Change in the Persian Gulf—Regional Security, Sustainability Strategies and Research Needs', Conference on Climate Change, Social Stress and Violent Conflict, Hamburg, 19–20 November 2009.

INDEX